Lotus Illustrated DICTIONARY of ECOLOGY

Compiled and Edited by:
Diana Richards

Lotus PRESS
4263 / 3, Ansari Road,
Daryaganj, New Delhi-02

Lotus Illustrated DICTIONARY of ECOLOGY

ISBN 81 89093 27 4

Published by:
LOTUS PRESS
4263/3, Ansari Road, Daryaganj,
New Delhi-110002
Ph: 32903912, 23280047
E-mail: lotus_press@sify.com
www: lotuspress.co.in

Printed at: ***Chetna Printers,*** Delhi

Preface

Ecology is a subject we all happen to interact with, but are unaware of various facets of it. This dictionary of Ecology is developed keeping this very fact in mind. This dictionary of Ecology will provide everything, which a conscious reader must posses. Apart from providing precise definitions of various terms related to Ecology, it also provides various facets of Ecology and environment, which will create awareness among the readers. And this will certainly be beneficial for all the elements of the Ecology, and we are one of those elements. This dictionary consists of more than 2000 researched terms, their definitions and pictorial representations of many terms. Definitions are included from all sorts of fields associated with Ecology. Infact this dictionary is not only meant for students and teachers of Ecology, but also for the people who are very much concerned with the well-being of our ecosystem, as it contains terms which are commonly used pertaining to the ecological system.

This dictionary is of a different kind from the other dictionaries already in the bookstores because of the fact that it provides information of Ecology in a different manner, which nobody has adopted till now. This dictionary provides rich, exhaustive and easy-to-understand terminology and their explanations apart from providing pictorial representations. The pictorial representations are not just concerned with the related terms but provide extra information associated with that particular term, in providing supplementary knowledge and hence increasing the knowledge by leaps and bounds. Though it is not possible to include each and every term associated with Ecology but effort has been made to include the terms, which are more relevant to the subject and will help in the understanding of ecosystem find their place in the list. This dictionary also keeps a tab on the recent development in the ecosystem and hence apprises the readers with the latest information, and the final result is a wonderful dictionary of Ecology that is going to make a mark in the minds of readers.

7Q10
seven-day, consecutive low flow with a ten year return frequency; the lowest stream flow for seven consecutive days that would be expected to occur once in ten years.

abandoned well
a well whose use has been permanently discontinued or which is in a state of such disrepair that it cannot be used for its intended purpose.

abatement
reducing the degree or intensity of, or eliminating, pollution.

abatement debris
waste from remediation activities.

abiotic
nonliving or not containing any living organisms.

abiotic factors
environmental influences produced other than by living organisms; for example, temperature, wind patterns, humidity, pH, substrate rock type, and other physical and chemical influences.

absolute poverty
the lack of sufficient income in cash or exchange items for meeting the most basic needs of food, clothing, and shelter.

absorbed dose
in exposure assessment, the amount of a substance that penetrates an exposed organism's absorption barriers (e.g. skin, lung tissue, gastrointestinal tract) through physical or biological processes. The term is synonymous with internal dose.

absorption
the taking in of water and dissolved minerals and nutrients across cell membranes. Contrast with ingestion.

absorption barrier
any of the exchange sites of the body that permit uptake of various substances at different rates (e.g. skin, lung tissue, and gastrointestinal-tract wall).

accession
a sample of a crop variety collected at a specific location and time; may be of any size.

■ accident site
the location of an unexpected occurrence, failure or loss, either at a plant or along a transportation route, resulting in a release of hazardous materials.

■ acclimatisation
the physiological and behavioural adjustments of an organism to changes in its environment.

■ acid
a solution that has a pH less than 7; also a solution having this pH value.

■ acid aerosol
acidic liquid or solid particles small enough to become airborne. High concentrations can irritate the lungs and have been associated with respiratory diseases like asthma.

■ acid deposition
a complex chemical and atmospheric phenomenon that occurs when emissions of sulphur and nitrogen compounds and other substances are transformed by chemical processes in the atmosphere, often far from the original sources, and then deposited on earth in either wet or dry form. The wet forms, popularly called "acid rain," can fall to earth as rain, snow, or fog. The dry forms are acidic gases or particulates.

■ acid fallout
molecules of acid formed from reactions high in the atmosphere involving nitrogen, sulphur oxides, and water vapour that settle out of the atmosphere without any additional water.

■ acid mine drainage
drainage of water from areas that have been mined for coal or other mineral ores. The water has a low pH because of its contact with sulphur-bearing material and is harmful to aquatic organisms.

■ acid neutralising capacity
measure of ability of a base (e.g. water or soil) to resist changes in pH.

■ acid precipitation
includes acid rain, acid fog, acid snow, and any other form of precipitation that is more acidic that normal (i.e., less that pH 5.6). Excess acidity

is derived from certain air pollutants namely sulphur dioxide and oxides of nitrogen. The effects can include fish kills and eutrophication of lakes; tree kills, leading to soil erosion; and physical corrosive damage to vehicles and buildings. Many historic buildings in India are suffering damage from severe corrosion due to acid precipitation.

■ **acid rain**

any kind of precipitation that is acid.

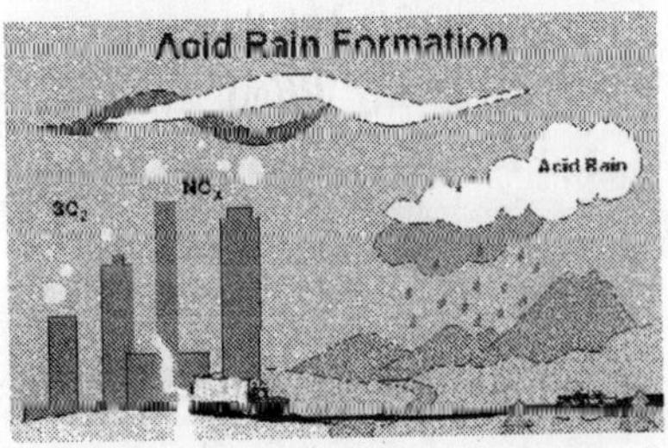

■ **acidic**

the condition of water or soil that contains a sufficient amount of acid substances to lower the pH below 7.0.

■ **activated carbon**

a highly adsorbent form of carbon used to remove odours and toxic substances from liquid or gaseous emissions. In waste treatment, it is used to remove dissolved organic matter from waste drinking water. It is also used in motor vehicle evapourative control systems.

■ **activated sludge**

product that results when primary effluent is mixed with bacteria-laden sludge and then agitated and aerated to promote biological treatment, speeding the breakdown of organic matter in raw sewage undergoing secondary waste treatment.

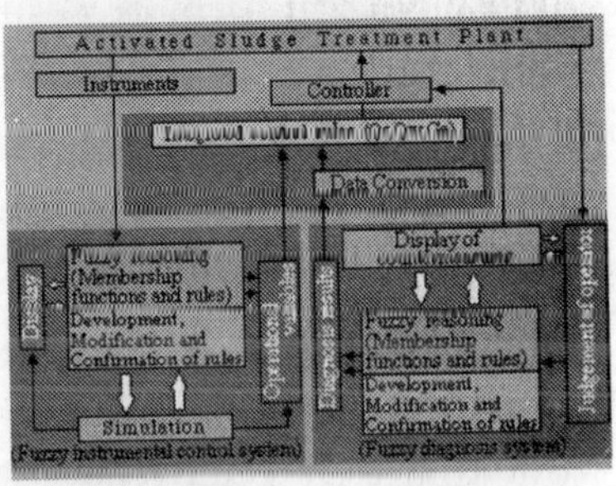

■ **activator**

a chemical added to a pesticide to increase its activity.

active ingredient
in any pesticide product, the component that kills, or otherwise controls, target pests. Pesticides are regulated primarily on the basis of active ingredients.

activity plans
written procedures in a school's asbestos-management plan that detail the steps a Local Education Agency (LEA) will follow in performing the initial and additional cleaning, operation and maintenance-program tasks; periodic surveillance; and re-inspection required by the Asbestos Hazard Emergency Response Act (AHERA).

acute exposure
a single exposure to a toxic substance, which may result in severe biological harm or death. Acute exposures are usually Characterised as lasting no longer than a day, as compared to longer, continuing exposure over a period of time.

acute toxicity
the ability of a substance to cause severe biological harm or death soon after a single exposure or dose. Also, any poisonous effect resulting from a single short-term exposure to a toxic substance.

adaptation
changes in an organism's physiological structure or function or habits that allow it to survive in new surroundings.

add-on control device
an air pollution control device such as carbon absorber or incinerator that reduces the pollution in an exhaust gas. The control device usually does not affect the process being controlled and thus is "add-on" technology, as opposed to a scheme to control pollution through altering the basic process itself.

adequately wet
asbestos containing material that is sufficiently mixed or penetrated with liquid to prevent the release of particulates.

administered dose
in exposure assessment, the amount of a substance given to a test subject (human or animal) to determine dose-response relationships. Since

exposure to chemicals is usually inadvertent, this quantity is often called potential dose.

■ **adsorption**

removal of a pollutant from air or water by collecting the pollutant on the surface of a solid material; e.g., an advanced method of treating waste in which activated carbon removes organic matter from waste-water.

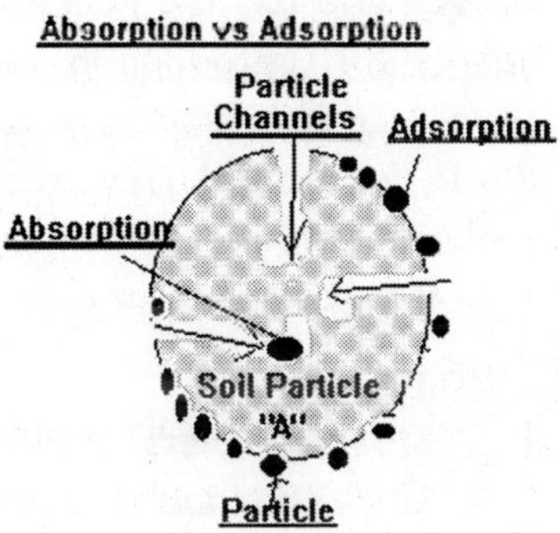

■ **adulterants**

chemical impurities or substances that by law do not belong in a food, or pesticide.

■ **adulterated**

1. any pesticide whose strength or purity falls below the quality stated on its label.
2. a food, feed, or product that contains illegal pesticide residues.

■ **advanced treatment**

a level of wastewater treatment more stringent than secondary treatment; requires an 85-percent reduction in conventional pollutant concentration or a significant reduction in non-conventional pollutants. Sometimes called tertiary treatment.

■ **advanced wastewater treatment**

any treatment of sewage that goes beyond the secondary or biological water treatment stage and includes the removal of nutrients such as phosphorus and nitrogen and a high percentage of suspended solids.

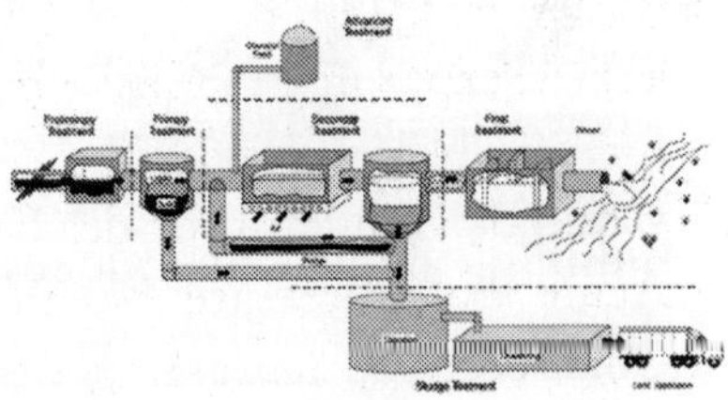

■ **advisory**

a non-regulatory document that communicates risk information to those who may have to make risk management decisions.

■ aerated lagoon

a holding and/or treatment pond that speeds up the natural process of biological decomposition of organic waste by stimulating the growth and activity of bacteria that degrade organic waste.

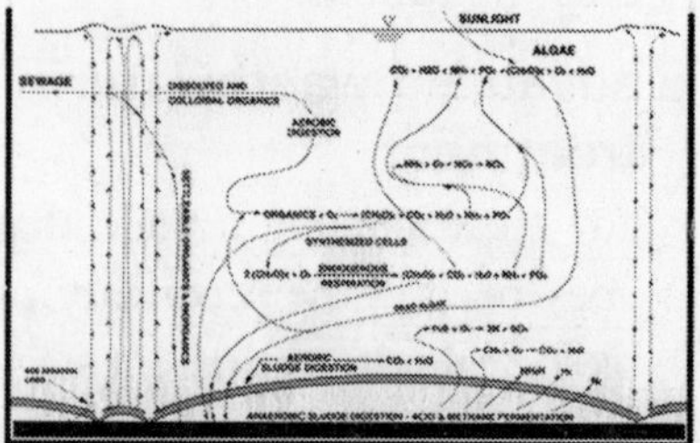

BIOLOGICAL REACTIONS IN AERATED-FACULTATIVE LAGOON PROCESS

■ aeration

a process, which promotes biological degradation of organic matter in water. The process may be passive (as when waste is exposed to air), or active (as when a mixing or bubbling device introduces the air).

■ aeration tank

a chamber used to inject air into water.

■ aerobe

an organism that utilises atmospheric oxygen (O_2) in its metabolic pathways. An organism that must have oxygen in order to survive is an obligate aerobe .

■ aerobic

pertaining to the presence of free oxygen. Aerobic organisms require oxygen for their life processes.

■ aerobic treatment

process by which microbes decompose complex organic compounds in the presence of oxygen and use the liberated energy for reproduction and growth. (Such processes include extended aeration, trickling filtration, and rotating biological contactors.).

■ aerosol

1. small droplets or particles suspended in the atmosphere, typically containing sulphur. They are usually emitted naturally (e.g. in volcanic eruptions) and as the result of anthropogenic (human) activities such as burning fossil fuels. 2. The pressurised gas used to propel substances out of a container.

■ affected public

1. The people who live and/or work near a hazardous waste site. 2. The human

population adversely impacted following exposure to a toxic pollutant in food, water, air, or soil.

■ afterburner

in incinerator technology, a burner located so that the combustion gases are made to pass through its flame in order to remove smoke and odours. It may be attached to or be separated from the incinerator proper.

■ age tank

a tank used to store a chemical solution of known concentration for feed to a chemical feeder. Also called a day tank.

■ agent

any physical, chemical, or biological entity that can be harmful to an organism (synonymous with stressors.).

■ agent orange

a toxic herbicide and defoliant used in the Vietnam conflict, containing 2,4,5-trichlorophen-oxyacetic acid (2,4,5-T) and 2-4 dichlorophenoxyacetic acid (2,4-D) with trace amounts of dioxin.

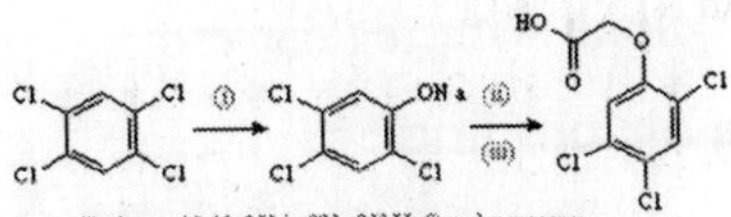

(i) heat with NaOH in CH_3OH/H_2O under pressure
(ii) $ClCH_2CO_2H$ in NaOH at 140°C
(iii) H^+

■ agricultural pollution

farming wastes, including runoff and leaching of pesticides and fertilise rs; erosion and dust from Ploughing; improper disposal of animal manure and carcasses; crop residues, and debris.

■ agriculture

the science of producing healthy plants and animals for food and other uses.

■ agriculturist

an expert in agriculture, such as a farmer or rancher.

■ agroecosystem

land used for crops, pasture, and livestock; the adjacent uncultivated land that supports other vegetation and wildlife; and the associated atmosphere, the underlying soils, groundwater, and drainage networks.

■ agroforestry

production of tree crops in a manner similar to agriculture.

Also, production of trees along with regular crops.

air binding

situation where air enters the filter media and harms both the filtration and backwash processes.

air changes per hour

the movement of a volume of air in a given period of time; if a house has one air change per hour, it means that the air in the house will be replaced in a one-hour period.

air cleaning

indoor-air quality-control strategy to remove various airborne particulates and/or gases from the air. Most common methods are particulate filtration, electrostatic precipitation, and gas sorption.

air contaminant

any particulate matter, gas, or combination thereof, other than water vapour.

air curtain

a method of containing oil spills. Air bubbling through a perforated pipe causes an upward water flow that slows the spread of oil. It can also be used to stop fish from entering polluted water.

air exchange rate

the rate at which outside air replaces indoor air in a given space.

air gap

open vertical gap or empty space that separates drinking water supply to be protected from another water system in a treatment plant or other location. The open gap protects the drinking water from contamination by backflow or back siphonage.

air handling unit

equipment that includes a fan or blower, heating and/or cooling coils, regulator controls, condensate drain pans, and air filters.

air mass

a large volume of air with certain meteorological or polluted characteristics—e.g., a heat inversion or smogginess while in one location. The characteristics can change as the air mass moves away.

air monitoring

see **monitoring.**

■ air padding
pumping dry air into a container to assist with the withdrawal of liquid or to force a liquefied gas such as chlorine out of the container.

■ air permeability
permeability of soil with respect to air. Important to the design of soil-gas surveys. Measured in darcys or centimeters-per-second.

■ air plenum
any space used to convey air in a building, furnace, or structure. The space above a suspended ceiling is often used as an air plenum.

■ air pollutant
any substance in air that could, in high enough concentration, harm man, other animals, vegetation, or material. Pollutants may include almost any natural or artificial composition of airborne matter capable of being airborne. They may be in the form of solid particles, liquid droplets, gases, or in combination thereof. Generally, they fall into two main groups (1) those emitted directly from identifiable sources and (2) those produced in the air by interaction between two or more primary pollutants, or by reaction with normal atmospheric constituents, with or without photo activation. Exclusive of pollen, fog, and dust, which are of natural origin, about 100 contaminants have been identified. Air pollutants are often grouped in categories for ease in classification; some of he categories are solids, sulphur compounds, volatile organic chemicals, particulate matter, nitrogen compounds, oxygen compounds, halogen compounds, radioactive compound, and odours.

■ air pollution
the presence of contaminants or pollutant substances in the air that interfere with human health or welfare, or produce other harmful environmental effects.

■ air pollution control device
mechanism or equipment that cleans emissions generated by a source (e.g. an incinerator, industrial smokestack, or an automobile exhaust system) by removing pollutants that

would otherwise be released to the atmosphere.

■ **air pollution episode**
a period of abnormally high concentration of air pollutants, often due to low winds and temperature inversion, that can cause illness and death.

■ **air quality criteria**
the levels of pollution and lengths of exposure above which adverse health and welfare effects may occur.

■ **air quality standards**
the level of pollutants prescribed by regulations that are not be exceeded during a given time in a defined area.

■ **air sparging**
injecting air or oxygen into an aquifer to strip or flush volatile contaminants as air bubbles up through The ground water and is captured by a vapour extraction system.

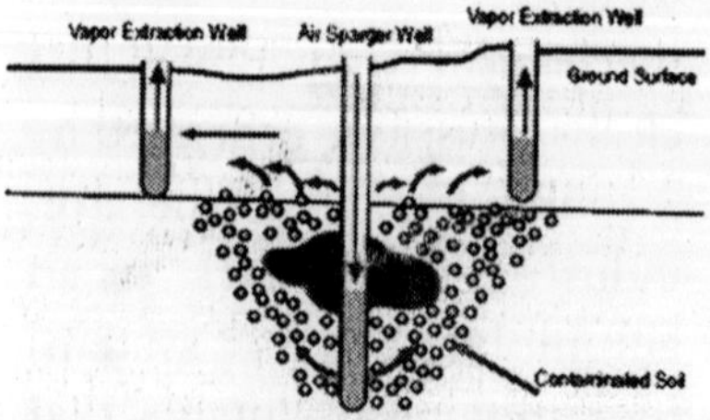

■ **air stripping**
a treatment system that removes volatile organic compounds (VOCs) from contaminated ground water or surface water by forcing an airstream through the water and causing the compounds to evapourate.

■ **air toxics**
any air pollutant for which a national ambient air quality standard (NAAQS) does not exist (i.e. excluding ozone, carbon monoxide, PM-10, sulphur dioxide, nitrogen oxide) that may reasonably be anticipated to cause cancer; respiratory, cardiovascular, or developmental effects; reproductive dysfunctions, neurological disorders, heritable gene mutations, or other serious or irreversible chronic or acute health effects in humans.

■ **air/oil table**
the surface between the vadose zone and ambient oil; the pressure of oil in the porous medium is equal to atmospheric pressure.

■ airborne particulates
total suspended particulate matter found in the atmosphere as solid particles or liquid droplets. Chemical composition of particulates varies widely, depending on location and time of year. Sources of airborne particulates include dust, emissions from industrial processes, combustion products from the burning of wood and coal, combustion products associated with motor vehicle or non-road engine exhausts, and reactions to gases in the atmosphere.

■ airborne release
release of any pollutant into the air.

■ alachlor
a herbicide, marketed under the trade name Lasso, used mainly to control weeds in corn and soybean fields.

■ alar
trade name for daminozide, a pesticide that makes apples redder, firmer, and less likely to drop off trees before growers are ready to pick them. It is also used to a lesser extent on peanuts, tart cherries, concord grapes, and other fruits.

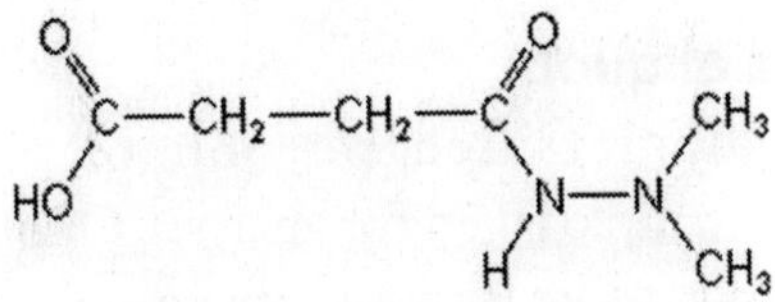

■ aldicarb
an insecticide sold under the trade name Temik. It is made from ethyl isocyanate.

■ algae
simple plantlike organisms found in water or on wet surfaces.

■ algal blooms
sudden spurts of algal growth, which can affect water quality adversely and indicate potentially hazardous changes in local water chemistry.

■ algicide
substance or chemical used specifically to kill or control algae.

■ alien species
a species occurring in an area outside of its historically known natural range as a result of intentional or acciden-

tal dispersal by human activities. Also known as introduced species

aliquot

a measured portion of a sample taken for analysis. One or more aliquots make up a sample. (See duplicate.).

alkaline

the condition of water or soil which contains a sufficient amount of alkali substance to raise the pH above 7.0.

alkalinity

the capacity of bases to neutralise acids. An example is lime added to lakes to decrease acidity.

allele

one of several forms of the same gene.

allergen

a substance that causes an allergic reaction in individuals sensitive to it.

alluvial

relating to and/or sand deposited by flowing water.

alternative compliance

a policy that allows facilities to choose among methods for achieving emission-reduction or risk-reduction instead of command-and control regulations that specify standards and how to meet them. Use of a theoretical emissions bubble over a facility to cap the amount of pollution emitted while allowing the company to choose where and how (within the facility) it complies.

alternative fuels

substitutes for traditional liquid, oil-derived motor vehicle fuels like gasoline and diesel. Includes mixtures of alcohol-based fuels with gasoline, methanol, ethanol, compressed natural gas, and others.

ambient air

any unconfined portion of the atmosphere open air, surrounding air.

ambient air quality standards

(See Criteria Pollutants and National Ambient Air Quality Standards.).

ambient measurement

a measurement of the concentration of a substance or pollutant within the immediate environs of an organism;

taken to relate it to the amount of possible exposure.

■ ambient medium

material surrounding or contacting an organism (e.g. outdoor air, indoor air, water, or soil, through which chemicals or pollutants can reach the organism. (See biological medium, environmental medium.).

■ ambient temperature

temperature of the surrounding air or other medium.

■ amprometric titration

a way of measuring concentrations of certain substances in water using an electric current that flows during a chemical reaction.

■ anaerobe

1. An organism capable of living in the absence of free oxygen (O_2). 2 Obligate anaerobe An organism that must live without oxygen, for whom oxygen (O_2) is toxic.

■ anaerobic decomposition

reduction of the net energy level and change in chemical composition of organic matter caused by microorganisms in an oxygen-free environment.

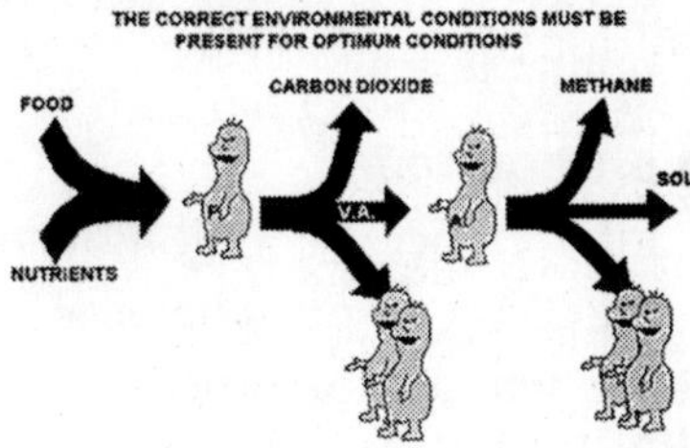

■ anaerobic

1. lacking or seriously depleted of oxygen. Opposite of aerobic. 2 Of or relating to organisms, such as certain bacteria, that can live in the absence of atmospheric oxygen (indeed, for most anaerobic bacteria, oxygen is toxic).

■ animal dander

tiny scales of animal skin, a common indoor air pollutant.

■ animal studies

investigations using animals as surrogates for humans with the expectation that the results are pertinent to humans.

■ anisotropy

in hydrology, the conditions under which one or more hydraulic properties of an aquifer vary from a reference point.

■ **annular space, annulus**
the space between two concentric tubes or casings, or between the casing and the borehole wall.

■ **antagonism**
interference or inhibition of the effect of one chemical by the action of another.

■ **antarctic "ozone hole"**
refers to the seasonal depletion of ozone in the upper atmosphere above a large area of Antarctica. (See Ozone Hole).

■ **antarctic circle**
the imaginary boundary of the southern polar region.

■ **anti-degradation clause**
part of federal air quality and water quality requirements prohibiting deterioration where pollution levels are above the legal limit.

■ **anti-microbial**
an agent that kills microbes.

■ **applied dose**
in exposure assessment, the amount of a substance in contact with the primary absorption boundaries of an organism (e.g. skin, lung tissue, gastrointestinal track) and available for absorption.

■ **aqueous**
something made up of water.

■ **aqueous solubility**
the maximum concentration of a chemical that will dissolve in pure water at a reference temperature.

■ **aquifer**
an underground geological formation, or group of formations, containing water. Are sources of groundwater for wells and springs.

■ **aquifer test**
a test to determine hydraulic properties of an aquifer.

■ **aquitard**
geological formation that may contain groundwater but is not capable of transmitting significant quantities of it under normal hydraulic gradients. May function as confining bed.

■ **architectural coatings**
coverings such as paint and roof tar that are used on exteriors of buildings.

■ arctic circle

the imaginary boundary of the northern polar region.

■ area source

any source of air pollution that is released over a relatively small area but which cannot be classified as a point source. Such sources may include vehicles and other small engines, small businesses and household activitics, or biogenic sources such as a forest that releases hydrocarbons.

■ aromatics

a type of hydrocarbon, such as benzene or toluene, with a specific type of ring structure. Aromatics are sometimes added to gasoline in order to increase octane. Some aromatics are toxic.

■ arsenicals

pesticides containing arsenic.

■ artesian (aquifer or well)

water held under pressure in porous rock or soil confined by impermeable geological formations.

■ arthropods

the animal phylum comprised of crustaceans, spiders, mites, centipedes, insects, and related forms. The largest of the phyla, containing more than three times the number of all other animal phyla combined.

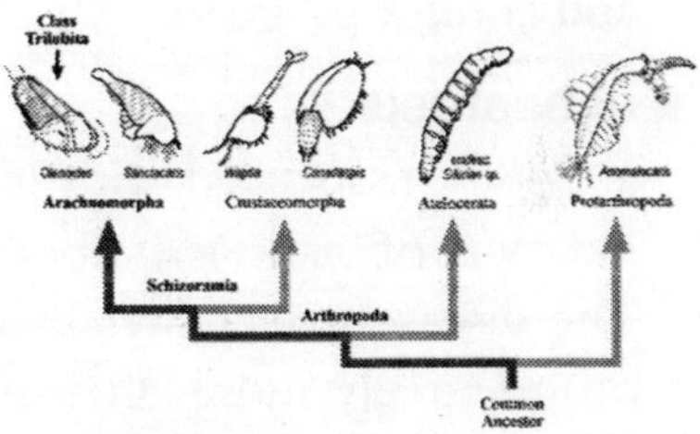

■ asbestos

a mineral fiber that can pollute air or water and cause cancer or asbestosis when inhaled.

■ asbestos abatement

procedures to control fiber release from asbestos-containing materials in a building or to remove them entirely, including removal, encapsulation, repair, enclosure, encasement, and operations and maintenance programs.

■ asbestos assessment

in the asbestos-in-schools program, the evaluation of the physical condition and potential for damage of all friable asbestos containing materials and thermal insulation systems.

■ **asbestos program manager**

a building owner or designated representative who supervises all aspects of the facility asbestos management and control program.

■ **asbestosis**

a disease associated with inhalation of asbestos fibers. The disease makes breathing progressively more difficult and can be fatal.

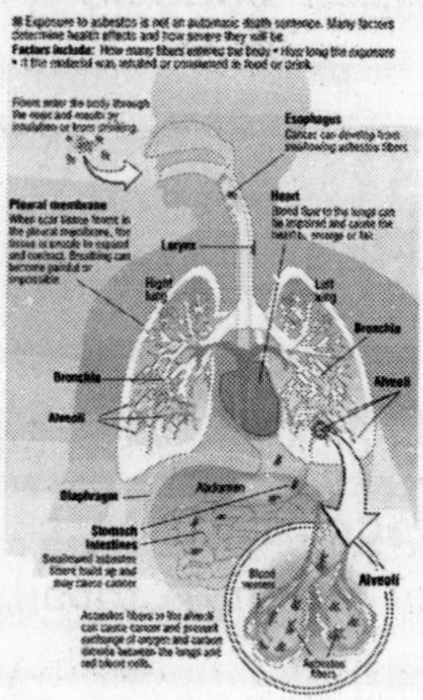

■ **a-scale sound level**

a measurement of sound approximating the sensitivity of the human ear, used to note the intensity or annoyance level of sounds.

■ **ash**

the mineral content of a product remaining after complete combustion.

■ **assay**

a test for a specific chemical, microbe, or effect.

■ **assemblage**

see "Community."

■ **assessment endpoint**

in ecological risk assessment, an explicit expression of the environmental value to be protected; includes both an ecological entity and specific attributed thereof. entity (e.g. salmon are a valued ecological entity; reproduction and population maintenance—the attribute—form an assessment endpoint.).

■ **assimilation**

the ability of a body of water to purify itself of pollutants.

■ **assimilative capacity**

the capacity of a natural body of water to receive wastewaters or toxic materials without deleterious effects and without damage to aquatic life or humans who consume the water.

association of boards of certification
an international organisation representing boards which certify the operators of waterworks and wastewater facilities.

atmosphere
the air in the universe.

atom
one of the tiny particles of which all things are made.

attainment area
an area considered to have air quality as good as or better than the national ambient air quality standards as defined in the Clean Air Act. An area may be an attainment area for one pollutant and a non-attainment area for others.

attenuation
the process by which a compound is reduced in concentration over time, through absorption, adsorption, degradation, dilution, and/or transformation. an also be the decrease with distance of sight caused by attenuation of light by particulate pollution.

attractant
a chemical or agent that lures insects or other pests by stimulating their sense of smell.

attrition
wearing or grinding down of a substance by friction. Dust from such processes contributes to air pollution.

autotroph
literally, "self eater." Organisms capable of producing their own food. Contrast with heterotroph .

available chlorine
a measure of the amount of chlorine available in chlorinated lime, hypochlorite compounds, and other materials used as a source of chlorine when compared with that of liquid or gaseous chlorines.

avifauna
all of the birds found in a given area.

avoided cost
the cost a utility would incur to generate the next increment of electric capacity using its own resources; many landfill gas projects' buy back

rates are based on avoided costs.

back pressure
a pressure that can cause water to backflow into the water supply when a user's waste water system is at a higher pressure than the public system.

backflow/back siphonage
a reverse flow condition created by a difference in water pressures that causes water to flow back into the distribution pipes of a drinking water supply from any source other than the intended one.

background extinction rate
normal rate of extinction as a natural part of the evolutionary process of various species as a result of changes in local environmental conditions and the actions of natural evolutionary forces. Extinctions not caused or contributed to by the actions of humans.

background level
1.the concentration of a substance in an environmental media (air, water, or soil) that occurs naturally or is not the result of human activities. 2. In exposure assessment the concentration of a substance in a defined control area, during a fixed period of time before, during, or after a data-gathering operation..

backwashing
reversing the flow of water back through the filter media to remove entrapped solids.

backyard composting
diversion of organic food waste and yard trimmings from the municipal waste stream by composting hem in one's yard through controlled decomposition of organic matter by bacteria and fungi into a humus-like product. It is considered source reduction, not recycling, because the composted materials never enter the municipal waste stream.

bacteria
(singular bacterium) microscopic living organisms that can aid in pollution control by metabolising organic matter in sewage, oil spills or other pollutants. However, bacteria

in soil, water or air can also cause human, animal and plant health problems.

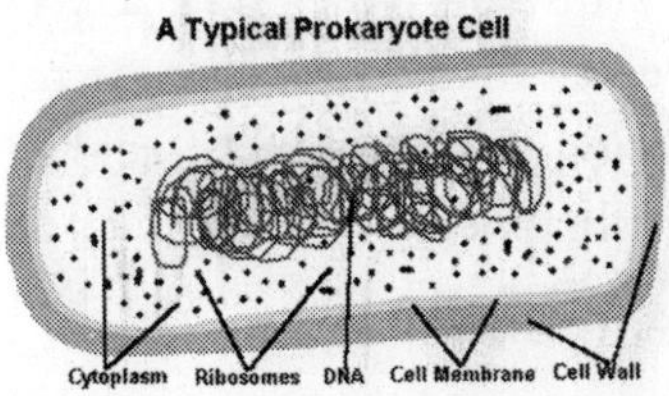

■ baffle

a flat board or plate, deflector, guide, or similar device constructed or placed in flowing water or slurry systems to cause more uniform flow velocities to absorb energy and to divert, guide, or agitate liquids.

■ baffle chamber

in incinerator design, a chamber designed to promote the settling of fly ash and coarse particulate matter by changing the direction and/or reducing the velocity of the gases produced by the combustion of the refuse or sludge.

■ baghouse filter

large fabric bag, usually made of glass fibers, used to eliminate intermediate and large (greater than 20 PM in diameter) particles. This device operates like the bag of an electric vacuum cleaner, passing the air and smaller particles while entrapping the larger ones.

■ bailer

a pipe with a valve at the lower end, used to remove slurry from the bottom or side of a well as it is being drilled, or to collect groundwater samples from wells or open boreholes. 2. A tube of varying length.

■ baling

compacting solid waste into blocks to reduce volume and simplify handling.

■ ballistic separator

a machine that sorts organic from inorganic matter for composting.

■ band application

the spreading of chemicals over, or next to, each row of plants in a field.

■ banking

a system for recording qualified air emission reductions for later use in bubble, offset, or netting transactions. (See emissions trading.).

■ bar screen

in wastewater treatment, a device used to remove large solids.

■ barrel sampler

open-ended steel tube used to collect soil samples.

■ barrier coating(s)

a layer of a material that obstructs or prevents passage of something through a surface that is to be protected; e.g., grout, caulk, or various sealing compounds; sometimes used with polyurethane membranes to prevent corrosion or oxidation of metal surfaces, chemical impacts on various materials, or, for example, to prevent radon infiltration through walls, cracks, or joints in a house.

■ basal application

in pesticides, the application of a chemical on plant stems or tree trunks just above the soil line.

■ basalt

consistent year-round energy use of a facility; also refers to the minimum amount of electricity supplied continually to a facility.

■ base

a solution that has a pH greater than 7; capable of reducing the amount of acid in a substance.

■ bean sheet

common term for a pesticide data package record.

■ bed load

sediment particles resting on or near the channel bottom that are pushed or rolled along by the flow of water.

■ bench-scale tests

laboratory testing of potential cleanup technologies (See treatability studies.).

■ benefit-cost analysis

an economic method for assessing the benefits and costs of achieving alternative health-based standards at given levels of health protection.

■ benthic

organisms that live on the bottom of the ocean are called benthic organisms. They are not free-floating like pelagic organisms are.

■ benthos

animals, such as clams, and plants, such as kelp, that live on the bottom of the ocean regardless of water depth. Most plants live in shallow water.

■ bentonite

a colloidal clay, expansible when moist, commonly used to provide a tight seal around a well casing.

■ beryllium

an metal hazardous to human health when inhaled as an airborne pollutant. It is discharged by machine shops, ceramic and propellant plants, and foundries.

■ bimetal

beverage containers with steel bodies and aluminium tops; handled differently from pure aluminium in recycling.

■ binomial nomenclature

the two-name system, developed by Carolus Linnaeus (the founder of modern taxonomy), used to assign scientific names to all living things. Homo sapiens, for example, is the scientific name for humans. The first name is the genus name and is always capitalised. This is sort of like your last name... it belongs to several of your close relatives, too, and it shows that you are all closely related. The second name is the species name is always lower case. This is like your first name, which no one else in your circle of relatives posseses and so it uniquely identifies you. Memory tool you probably know the meanings of the terms generic (i.e. general, broad) and specific (i.e. precise, exact). These terms come from the same origins as genus and species, so recalling their meaning will help you recall the relationship between the two portions of a scientific name.

■ **bioaccumulation**

an increase in the concentration of a chemical in specific organs or tissues at a level higher than would normally be expected.

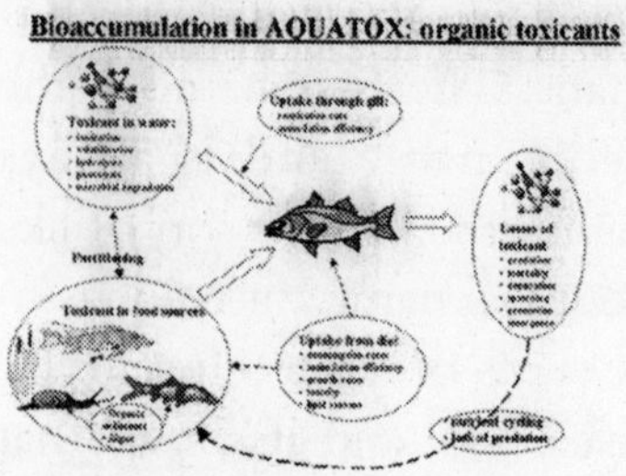

■ **bioassay**

a test to determine the relative strength of a substance by comparing its effect on a test organism with that of a standard preparation.

■ **bioavailabiliity**

degree of ability to be absorbed and ready to interact in organism metabolism.

■ **biochemical oxygen demand (BOD)**

a measure of the amount of oxygen consumed in the biological processes that break down organic matter in water. The greater the BOD, the greater the degree of pollution.

■ **bioconcentration**

the accumulation of a chemical in tissues of a fish or other organism to levels greater than in the surrounding medium.

■ **biodegradable**

able to be broken down into simpler substances (elements and compounds) by naturally occuring decomposers. Essentially, anything that can be ingested by an organism without causing that organism harm. 2. Nontoxic and able to be decomposed in relatively short period even on a human time scale .

■ **biodiversity**

the variety of biotic factors found within a specified geographic region. 2. The combined differences of living things, generally classified in four broad categories.

■ **biogeochemical cycles**

the movement of massive amounts of carbon, nitrogen, oxygen, hydrogen, calcium, sodium, sulphur, phosphorus, and other elements among various living and non-living components of the environment—including the atmosphere, soils, aquatic systems, and biotic systems—through the processes of production and decomposition.

■ **biogeography**
the scientific study of the geographic distribution of organisms.

■ **biological contaminants**
living organisms or derivates (e.g. viruses, bacteria, fungi, and mammal and bird antigens) that can cause harmful health effects when inhaled, swallowed, or otherwise taken into the body.

■ **biological control**
in pest control, the use of animals and organisms that eat or otherwise kill or out-compete pests.

■ **biological integrity**
the ability to support and maintain balanced, integrated, functionality in the natural habitat of a given region. Concept is applied primarily in drinking water management.

■ **biological magnification**
refers to the process whereby certain substances such as pesticides or heavy metals move up the food chain, work their way into rivers or lakes, and are eaten by aquatic organisms such as fish, which in turn are eaten by large birds, animals or humans. The substances become concentrated in tissues or internal organs as they move up the chain. (See bioaccumulants.).

■ **biological measurement**
a measurement taken in a biological medium. For exposure assessment, it is related to the measurement is taken to related it to the established internal dose of a compound.

■ **biological medium**
one of the major component of an organism; e.g. blood, fatty tissue, lymph nodes or breath, in which chemicals can be stored or transformed.

■ **biological oxidation**
decomposition of complex organic materials by microorganisms. Occurs in self-purification of water bodies and in activated sludge wastewater treatment.

biological oxygen demand

an indirect measure of the concentration of biologically degradable material present in organic wastes. It usually reflects the amount of oxygen consumed in five days by biological processes breaking down organic waste.

biological resources

those components of biodiversity of direct, indirect, or potential use to humanity.

biological stressors

organisms accidentally or intentionally dropped into habitats in which they do not evolve naturally; e.g. gypsy moths, Dutch elm disease, certain types of algae, and bacteria.

biological treatment

a treatment technology that uses bacteria to consume organic waste.

biologically effective dose

the amount of a deposited or absorbed compound reaching the cells or target sites where adverse effect occur, or where the chemical interacts with a membrane.

biologicals

vaccines, cultures and other preparations made from living organisms and their products, intended for use in diagnosing, immunising, or treating humans or animals, or in related research.

biomass

all of the living material in a given area; often refers to vegetation.

biomes

the world's major communities, classified according to the predominant vegetation and Characterised by adaptations of organisms to that particular environment.

biomonitoring

1.the use of living organisms to test the suitability of effluents for discharge into receiving waters and to test the quality of such waters downstream from the discharge. 2. analyses of blood, urine, tissues, etc. to measure chemical exposure in humans.

■ **bionomics**
see ecology .

■ **bioregion (bioregional planning)**
a territory defined by a combination of biological, social, and geographic criteria, rather than geopolitical considerations; generally, a system of related, interconnected ecosystems.

■ **bioremediation**
use of living organisms to clean up oil spills or remove other pollutants from soil, water, or wastewater; use of organisms such as non-harmful insects to remove agricultural pests or counteract diseases of trees, plants, and garden soil.

■ **biosensor**
analytical device comprising a biological recognition element (e.g. enzyme, receptor, DNA, antibody, or microorganism) in intimate contact with an electrochemical, optical, thermal, or acoustic signal transducer that together permit analyses of chemical properties or quantities. Shows potential development in some areas, including environmental monitoring.

■ **biosphere**
the portion of the earth and its atmosphere in which living organisms exist or that is capable of supporting life. 2. All of earth's ecosystems combined into one inclusive unit. Also called the "ecosphere." 3. The living organisms and their environment composing the biosphere.

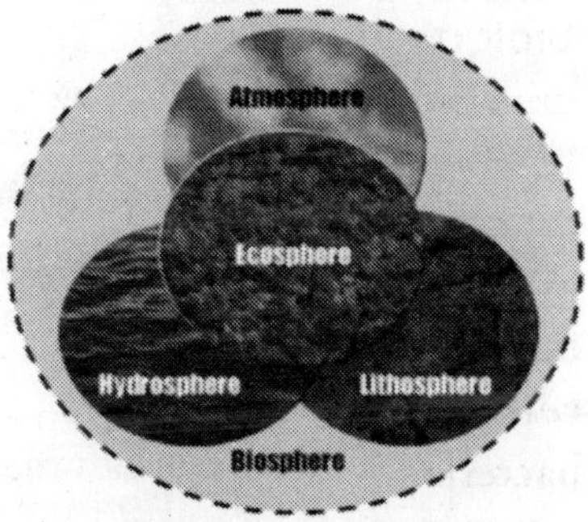

■ **biostabiliser**
a machine that converts solid waste into compost by grind ing and aeration.

■ **biota**
all of the organisms, including animals, plants, fungi, and microorganisms, found in a given area.

biotechnology
any technology that is applied to living organisms to make them more valuable to people.

biotic
pertaining to any aspect of life, especially to characteristics of entire populations or ecosystems.

biotic community
a naturally occurring assemblage of plants and animals that live in the same environment and are mutually sustaining and interdependent.

biotransformation
conversion of a substance into other compounds by organisms; includes biodegradation.

blackwater
water that contains animal, human, or food waste.

blood products
any product derived from human blood, including but not limited to blood plasma, platelets, red or white corpuscles, and derived licensed products such as interferon.

bloom
a proliferation of algae and/or higher aquatic plants in a body of water; often related to pollution, especially when pollutants accelerate growth.

BOD5
the amount of dissolved oxygen consumed in five days by biological processes breaking down organic matter.

body burden
the amount of a chemical stored in the body at a given time, especially a potential toxin in the body as the result of exposure.

bog
a type of wetland that accumulates appreciable peat deposits. Bogs depend primarily on precipitation for their water source, and are usually acidic and rich in plant residue with a conspicuous mat of living green moss.

boiler
a vessel designed to transfer heat produced by combustion or electric resistance to water. Boilers may provide hot water or steam.

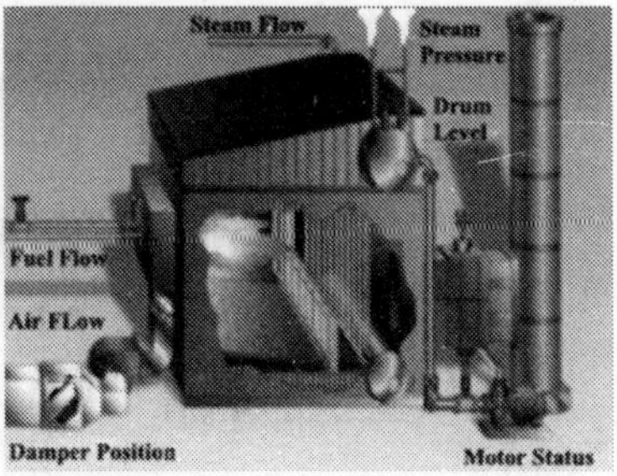

■ **boom**
1. A floating device used to contain oil on a body of water. 2 A piece of equipment used to apply pesticides from a tractor or truck.

■ **borehole**
hole made with drilling equipment.

■ **botanical pesticide**
a pesticide whose active ingredient is a plant-produced chemical such as nicotine or strychnine. Also called a plant-derived pesticide.

■ **bottom ash**
the non-airborne combustion residue from burning pulverised coal in a boiler; the material which falls to the bottom of the boiler and is removed mechanically; a concentration of non-combustible materials, which may include toxics.

■ **bottom land hardwoods**
forested freshwater wetlands adjacent to rivers in the southeastern United States, especially valuable for wildlife breeding, nesting and habitat.

■ **bounding estimate**
an estimate of exposure, dose, or risk that is higher than that incurred by the person in the population with the currently highest exposure, dose, or risk. Bounding estimates are useful in developing statements that exposures, doses, or risks are not greater than an estimated value.

■ **brackish**
mixed fresh and salt water.

■ **breakpoint chlorination**
addition of chlorine to water until the chlorine demand has been satisfied.

■ **breakthrough**
a crack or break in a filter bed that allows the passage of floc or particulate matter through a filter; will cause an increase in filter effluent turbidity.

■ **breathing zone**
area of air in which an organism inhales.

brine mud
waste material, often associated with well-drilling or mining, composed of mineral salts or other inorganic compounds.

british thermal unit
unit of heat energy equal to the amount of heat required to raise the temperature of one pound of water by one degree Fahrenheit at sea level.

broadcast application
the spreading of pesticides over an entire area.

brownfields
abandoned, idled, or under used industrial and commercial facilities/sites where expansion or redevelopment is complicated by real or perceived environmental contamination. They can be in urban, suburban, or rural areas.

bubble
a system under which existing emissions sources can propose alternate means to comply with a set of emissions limitations; under the bubble concept, sources can control more than required at one emission point where control costs are relatively low in return for a comparable relaxation of controls at a second emission point where costs are higher.

bubble policy
see **emissions trading.**

buffer
a solution or liquid whose chemical makeup is such that it Minimises changes in pH when acids or bases are added to it.

buffer strips
strips of grass or other erosion-resisting vegetation between or below cultivated strips or fields.

buffer zone
the region near the border of a protected area; a transition zone between areas managed for different objectives.

building cooling load
the hourly amount of heat that must be removed from a building to maintain indoor comfort (measured in British thermal units (BTUs).

building envelope
the exterior surface of a building's construction—the walls, windows, floors, roof,

and floor. Also called building shell.

■ **building related illness**
diagnosable illness whose cause and symptoms can be directly attributed to a specific pollutant source within a building (e.g. Legionnaire's disease, hypersensitivity, pneumonitis.).

■ **bulk sample**
a small portion (usually thumbnail size) of a suspect asbestos-containing building material collected by an asbestos inspector for laboratory analysis to determine asbestos content.

■ **bulky waste**
large items of waste materials, such as appliances, furniture, large auto parts, trees, stumps.

■ **burial ground (graveyard)**
a disposal site for radioactive waste materials that uses earth or water as a shield.

■ **buy-back center**
facility where individuals or groups bring recyclables in return for payment.

■ **by-product**
material, other than the principal product, generated as a consequence of an industrial process or as a breakdown product in a living system.

■ **cadmium (Cd)**
a heavy metal that accumulates in the environment.

■ **canopy layer**
a forest's second layer, or roof. Consists of a network of branches and leaves, forms a covering that blocks some of the sunlight from lower plants.

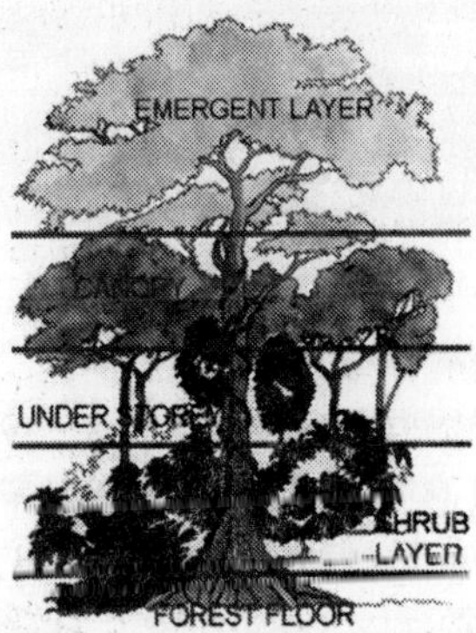

■ **cap**
a layer of clay, or other impermeable material installed over the top of a closed landfill to prevent entry of rainwater and Minimise leachate.

■ **capacity assurance plan**
a statewide plan which supports a state's ability to manage the hazardous waste generated within its boundaries over a twenty year period.

■ **capillary action**
movement of water through very small spaces due to molecular forces called capillary forces.

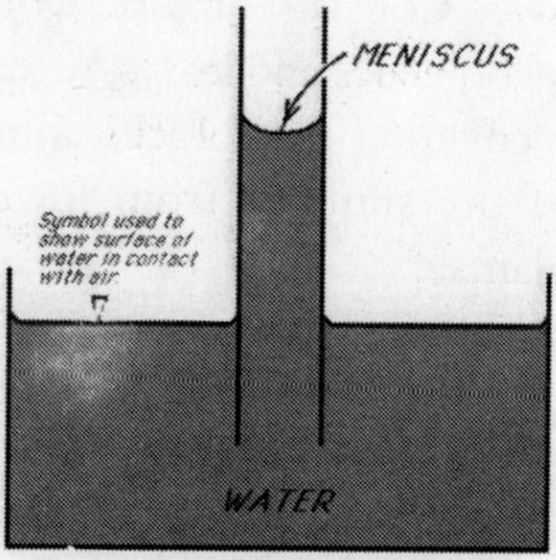

Water rising in a small glass tube by capillary action.

■ **capillary fringe**
the porous material just above the water table which may hold water by capillarity (a property of surface tension that draws water upwards) in the smaller void spaces.

■ **capture efficiency**
the fraction of organic vapours generated by a process that are directed to an abatement or recovery device.

■ **carbon absorber**
an add-on control device that uses activated carbon to absorb volatile organic compounds from a gas stream. (The VOCs are later recovered from the carbon.).

■ **carbon adsorption**
a treatment system that removes contaminants from ground water or surface water by forcing it through tanks containing activated carbon treated to attract the contaminants.

■ **carbon dioxide**
a gas found in air that is used in photosynthesis and produced by respiration; one of the green house gases.

■ **carbon monoxide (Co)**
a colourless, odourless, poisonous gas produced by incomplete fossil fuel combustion.

■ **carbon tetrachloride (CCl_4)**
compound consisting of one carbon atom ad four chlorine atoms, once widely used as a industrial raw material, as a solvent, and in the production of CFCs. Use as a solvent

ended when it was discovered to be carcinogenic.

■ **carboxyheamoglobin**

haemoglobin in which the iron is bound to carbon monoxide (CO) instead of oxygen.

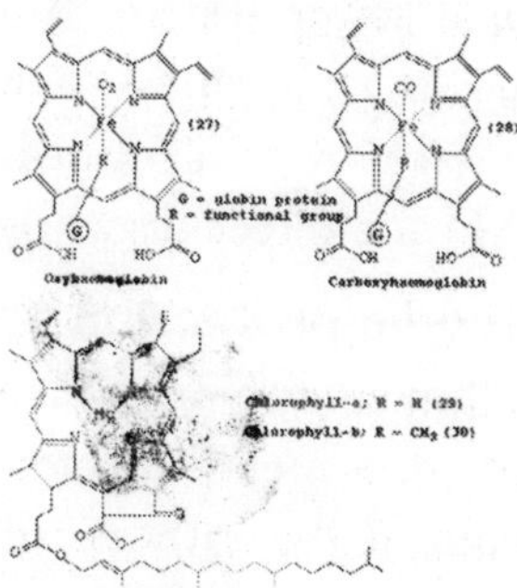

■ **carcinogen**

any substance that can cause or aggravate cancer.

■ **carnivores**

animals that eat other animals; see food chain

■ **carrier**

1. The inert liquid or solid material in a pesticide product that serves as a delivery vehicle for the active ingredient. Carriers do not have toxic properties of their own. 2. Any material or system that can facilitate the movement of a pollutant into the body or cells.

■ **carrying capacity**

the maximum number of people, or individuals of a particular species, that a given part of the environment can maintain indefinitely.

■ **cas registration number**

a number assigned by the Chemical Abstract Service to identify a chemical.

■ **case study**

a brief fact sheet providing risk, cost, and performance information on alternative methods and other pollution prevention ideas, compliance initiatives, voluntary efforts, etc.

■ **cask**

a thick-walled container (usually lead) used to transport radioactive material. Also called a coffin.

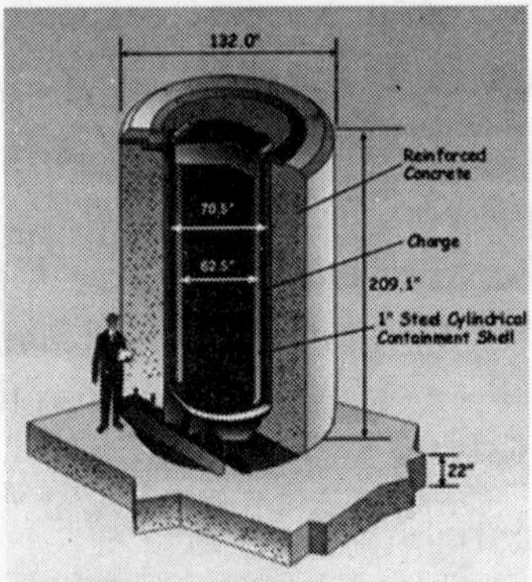

Spent Fuel Storage Cask

■ catalyst

a substance that changes the speed or yield of a chemical reaction without being consumed or chemically changed by the chemical reaction.

■ catalytic converter

an air pollution abatement device that removes pollutants from motor vehicle exhaust, either by oxidising them into carbon dioxide and water or reducing them to nitrogen.

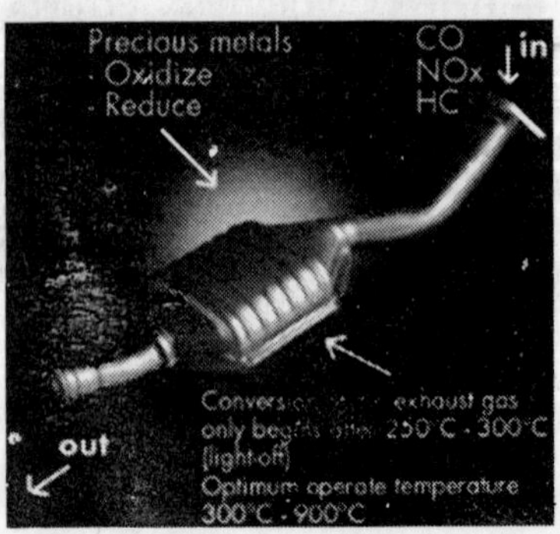

■ catalytic incinerator

a control device that oxidises volatile organic compounds (VOCs) by using a catalyst to promote the combustion process. Catalytic incinerators require lower temperatures than conventional thermal incinerators, thus saving fuel and other costs.

■ categorical pretreatment standard

a technology-based effluent limitation for an industrial facility discharging into a municipal sewer system. Analogous in stringency to Best Availability Technology (BAT) for direct dischargers.

■ cathodic protection

a technique to prevent corrosion of a metal surface by making it the cathode of an electrochemical cell.

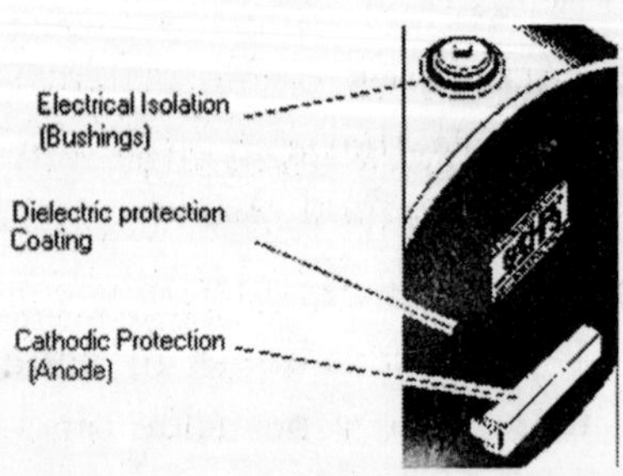

■ cavitation
the formation and collapse of gas pockets or bubbles on the blade of an impeller or the gate of a valve; collapse of these pockets or bubbles drives water with such force that it can cause pitting of the gate or valve surface.

■ cells
1. in solid waste disposal, holes where waste is dumped, compacted, and covered with layers of dirt on a daily basis. 2. the smallest structural part of living matter capable of functioning as an independent unit.

■ cellulose
plant fiber.

Section of cellulose showing bond between glucose molecules

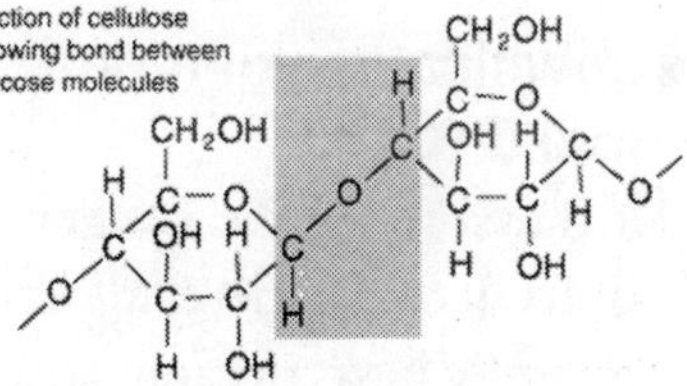

■ cementitious
densely packed and nonfibrous friable materials.

■ center of diversity
geographic region with high levels of genetic or species diversity.

■ center of endemism
geographic region with numerous locally endemic species.

■ central collection point
location were a generator of regulated medical waste consolidates wastes originally generated at various locations in his facility. The wastes are gathered together for treatment on-site or for transportation elsewhere for treatment and/or disposal. This term could also apply to community hazardous waste collections, industrial and other waste management systems.

■ centrifugal collector
a mechanical system using centrifugal force to remove aerosols from a gas stream or to remove water from sludge.

■ CFC's
chlorofluorocarbon gases that are released into the atmosphere as air pollutants and that change ozone to oxygen in the ozone layer.

■ channelisation
straightening and deepening streams so water will move

faster, a marsh-drainage tactic that can interfere with waste assimilation capacity, disturb fish and wildlife habitats, and aggravate flooding.

■ characterisation of ecological effects
part of ecological risk assessment that evaluates ability of a stressor to cause adverse effects under given circumstances.

■ characterisation of exposure
portion of an ecological risk assessment that evaluates interaction of a stressor with one or more ecological entities.

■ characteristic
any one of the four categories used in defining hazardous waste ignitability, corrosivity, reactivity, and toxicity.

■ characteristic diversity
the pattern of distribution and abundance of populations, species, and habitats under conditions where humanity's influence on the ecosystem is no greater than that of any other biotic factor.

■ check-valve tubing pump
water sampling tool also referred to as a water Pump.

■ chemical case
for purposes of review and regulation, the grouping of chemically similar pesticide active ingredients (e.g. salts and esters of the same chemical) into chemical cases.

■ chemical compound
a distinct and pure substance formed by the union or two or more elements in definite proportion by weight.

■ chemical element
a fundamental substance comprising one kind of atom; the simplest form of matter.

■ chemical oxygen demand (COD)
a measure of the oxygen required to oxidise all compounds, both organic and inorganic, in water.

■ chemical stressors
chemicals released to the environment through industrial waste, auto emissions, pesticides, and other human activity that can cause illnesses and

even death in plants and animals.

■ chemical treatment

any one of a variety of technologies that use chemicals or a variety of chemical processes to treat waste.

■ chemnet

mutual aid network of chemical shippers and contractors that assigns a contracted emergency response company to provide technical support if a representative of the firm whose chemicals are involved in an incident is not readily available.

■ chemosterilant

a chemical that controls pests by preventing reproduction.

■ chemtrec

the industry-sponsored Chemical Transportation Emergency Center; provides information and/or emer gency assistance to emergency responders.

■ child resistant packaging

packaging that protects children or adults from injury or illness resulting from accidental contact with or ingestion of residential pesticides that meet or exceed specific toxicity levels.

■ chiller

a device that generates a cold liquid that is circulated through an air-handling unit's cooling coil to cool the air supplied to the building.

■ chilling effect

the lowering of the Earth's temperature because of increased particles in the air blocking the sun's rays. (See greenhouse effect.).

■ chisel ploughing

preparing croplands by using a special implement that avoids complete inversion of the soil as in conventional Ploughing. Chisel Ploughing can leave a protective cover or crops residues on the soil surface to help prevent erosion and improve filtration.

■ chlorinated hydrocarbons

1.chemicals containing only chlorine, carbon, and hydrogen. These include a class of persistent, broad-spectrum insecticides that linger in the environment and accumulate in the food chain. Among them are DDT, aldrin, dieldrin, heptachlor, chlordane, lindane, endrin, Mirex, hexachloride, and toxaphene. Other examples include TCE, used as an industrial solvent. 2. Any chlorinated organic compounds including chlorinated solvents such as dichloromethane, trichloromethylene, chloroform.

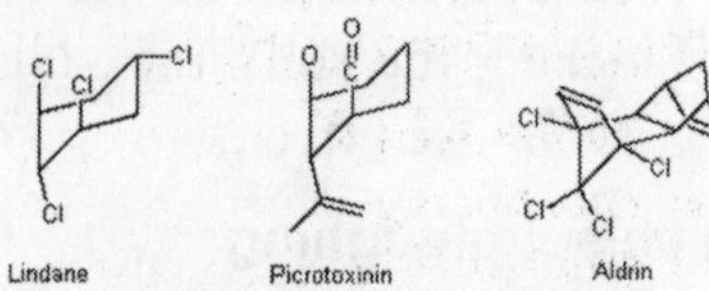

■ **chlorinated solvent**

an organic solvent containing chlorine atoms(e.g. methylene chloride and 1,1,1-trichloromethane). Uses of chlorinated solvents are include aerosol spray containers, in highway paint, and dry cleaning fluids.

■ **chlorination**

the application of chlorine to drinking water, sewage, or industrial waste to disinfect or to oxidise undesirable compounds.

■ **chlorinator**

a device that adds chlorine, in gas or liquid form, to water or sewage to kill infectious bacteria.

■ **chlorine-contact chamber**

that part of a water treatment plant where effluent is disinfected by chlorine.

■ **chlorophenoxy**

a class of herbicides that may be found in domestic water supplies and cause adverse health effects.

■ **chlorophyll**

a green light-absorbing pigment used in photosynthesis.

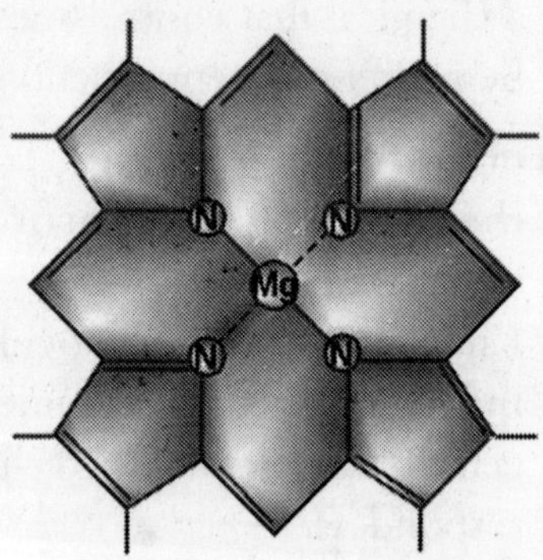

■ **chlorosis**

discolouration of normally green plant parts caused by disease, lack of nutrients, or various air pollutants.

■ **cholinesterase**
an enzyme found in animals that regulates nerve impulses by the inhibition of acetylcholine. Cholinesterase inhibition is associated with a variety of acute symptoms such as nausea, vomiting, blurred vision, stomach cramps, and rapid heart rate.

■ **chromium**
see **heavy metals.**

■ **chronic effect**
an adverse effect on a human or animal in which symptoms recur frequently or develop slowly over a long period of time.

■ **chronic exposure**
multiple exposures occurring over an extended period of time or over a significant fraction of an animal's or human's lifetime (Usually seven years to a lifetime.).

■ **chronic toxicity**
the capacity of a substance to cause long-term poisonous health effects in humans, animals, fish, and other organisms. (See acute toxicity.).

■ **circle of influence**
the circular outer edge of a depression produced in the water table by the pumping of water from a well. (See cone of depression.).

■ **cirrus**
thin, wispy clouds which form at the upper layer of the atmosphere and are composed mostly of ice crystals.

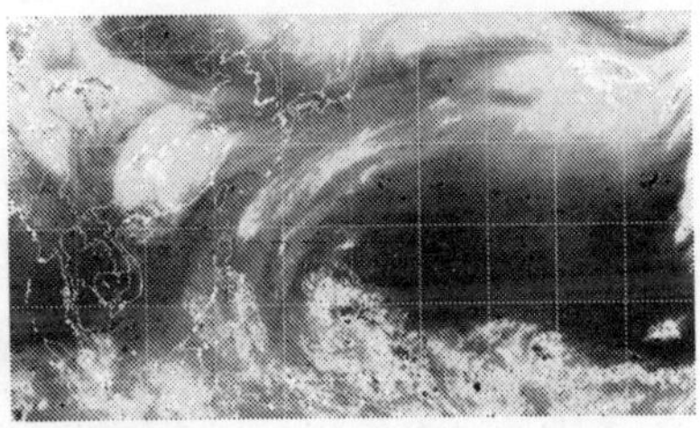

■ **cistern**
small tank or storage facility used to store water for a home or farm; often used to store rainwater.

■ **clarification**
clearing action that occurs during wastewater treatment when solids settle out. This is

often aided by centrifugal action and chemically induced coagulation in wastewater.

■ clarifier
a tank in which solids settle to the bottom and are subsequently removed as sludge.

■ class
in taxonomy, a category just beneath the phylum and above the order; a group of related, similar orders.

■ class 2 substance
a substance with an ozone depletion potential of less than 0.2. All HCFCs are currently included in this classification.

■ clay soil
soil material containing more than 40 percent clay, less than 45 percent sand, and less than 40 percent silt.

■ clean coal technology
any technology not in widespread use prior to the Clean Air Act Amendments of 1990. This Act will achieve significant reductions in pollutants associated with the burning of coal.

■ clean fuels
blends or substitutes for gasoline fuels, including compressed natural gas, methanol, ethanol, and liquified petroleum gas.

■ cleaner technologies substitutes assessment
a document that systematically evaluates the relative risk, performance, and cost trade-offs of technological alternatives; serves as a repository for all the technical data (including methodology and results) developed by a DfE or other pollution prevention or education project.

■ cleanup
actions taken to deal with a release or threat of release of a hazardous substance that could affect humans and/or the environment. The term "cleanup" is sometimes used interchangeably with the terms remedial action, removal action, response action, or corrective action.

■ clear cut
harvesting all the trees in one area at one time, a practice that can encourage fast rainfall or snowmelt runoff, ero-

sion, sedimentation of streams and lakes, and flooding, and destroys vital habitat.

■ clear well

a reservoir for storing filtered water of sufficient quantity to prevent the need to vary the filtration rate with variations in demand. Also used to provide chlorine contact time for disinfection.

■ climate

an area's weather over an extended period of time.

■ climate change

the term 'climate change' is sometimes used to refer to all forms of climatic inconsistency, but because the Earth's climate is never static, the term is more properly used to imply a significant change from one climatic condition to another. In some cases, 'climate change' has been used synonymously with the term, 'global warming'; scientists however, tend to use the term in the wider sense to also include natural changes in climate. (See global warming.).

■ climax community

the end of a successional sequence; a community that has reached stability under a particular set of environmental conditions.

■ cline

change in population characteristics over a geographical area, usually related to a corresponding environmental change.

■ clone

a population of individuals all derived asexually from the same single parent.

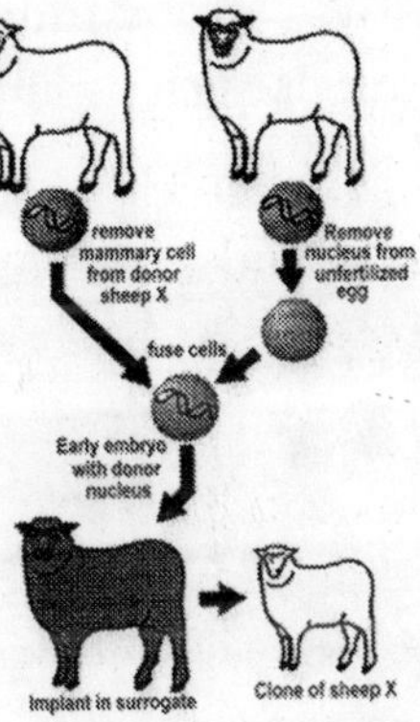

■ cloning

in biotechnology, obtaining a group of genetically identical cells from a single cell; making identical copies of a gene.

■ closed-loop recycling

reclaiming or reusing wastewater for non-potable purposes in an enclosed process.

■ **closure**

the procedure a landfill operator must follow when a landfill reaches its legal capacity for solid ceasing acceptance of solid waste and placing a cap on the landfill site.

■ **cloud**

visible water vapour suspended in the air.

■ **coagulation**

clumping of particles in wastewater to settle out impurities, often induced by chemicals such as lime, alum, and iron salts.

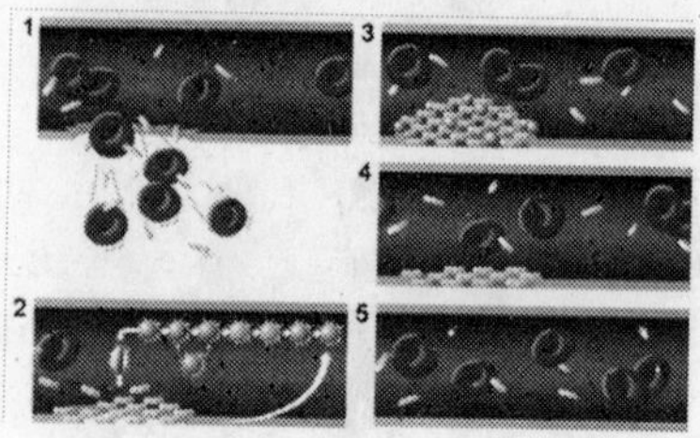

■ **coal cleaning technology**

a pre combustion process by which coal is physically or chemically treated to remove some of its sulphur so as to reduce sulphur dioxide emissions.

■ **coal gasification**

conversion of coal to a gaseous product by one of several available technologies.

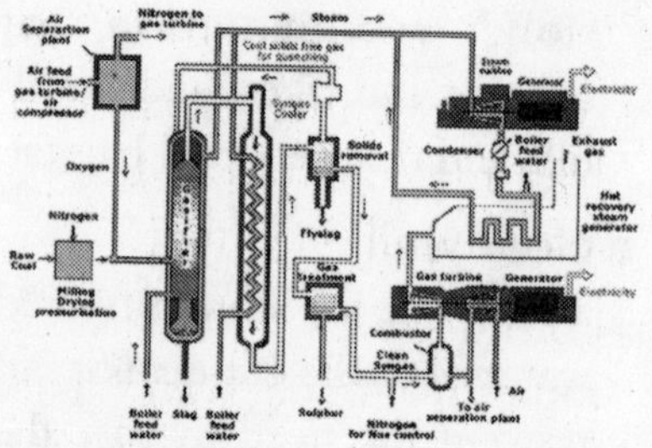

■ **coastal zone**

lands and waters adjacent to the coast that exert an influence on the uses of the sea and its ecology, or whose uses and ecology are affected by the sea.

■ **coefficient of haze**

a measurement of visibility interference in the atmosphere.

■ **co-fire**

burning of two fuels in the same combustion unit; e.g., coal and natural gas, or oil and coal.

■ **cogeneration**

the consecutive generation of useful thermal and electric energy from the same fuel source.

■ **coke oven**

an industrial process which converts coal into coke, one

of the basic materials used in blast furnaces for the conversion of iron ore into iron.

■ **cold desert**
a desert with daytime temperatures below freezing for part of the year.

■ **cold temperature, CO**
a standard for automobile emissions of carbon monoxide (CO) emissions to be met at a low temperature (i.e. 20 degrees Fahrenheit). Conventional automobile catalytic converters are not efficient in cold weather until they warm up.

■ **coliform index**
a rating of the purity of water based on a count of faecal bacteria.

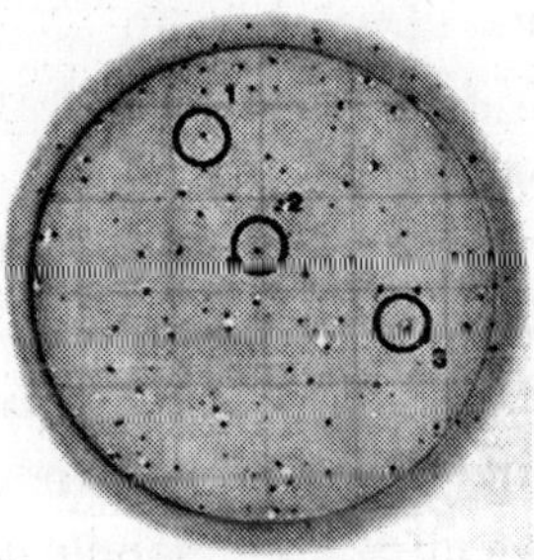

■ **coliform organism**
microorganisms found in the intestinal tract of humans and animals. Their presence in water indicates faecal pollution and potentially adverse contamination by pathogens.

■ **collector**
public or private hauler that collects non-hazardous waste and recyclable materials from residential, commercial, institutional and industrial sources. (See hauler.).

■ **collector sewers**
pipes used to collect and carry wastewater from individual sources to an interceptor sewer that will carry it to a treatment facility.

■ **colloids**
very small, finely divided solids (that do not dissolve) that remain dispersed in a liquid for a long time due to their small size and electrical charge.

■ **colonial animals**
animals that live in colonies.

■ **colony**
a large population whose members depend on each other.

■ **co-management**
the sharing of authority, responsibility, and benefits between government and local

communities in the management of natural resources.

■ **combined sewer overflows**

discharge of a mixture of storm water and domestic waste when the flow capacity of a sewer system is exceeded during rainstorms.

■ **combined sewers**

a sewer system that carries both sewage and storm-water runoff. Normally, its entire flow goes to a waste treatment plant, but during a heavy storm, the volume of water may be so great as to cause overflows of untreated mixtures of storm water and sewage into receiving waters. Storm-water runoff may also carry toxic chemicals from industrial areas or streets into the sewer system.

■ **combustion**

1. burning, or rapid oxidation, accompanied by release of energy in the form of heat and light. 2. Refers to controlled burning of waste, in which heat chemically alters organic compounds, converting into stable inorganic such as carbon dioxide and water.

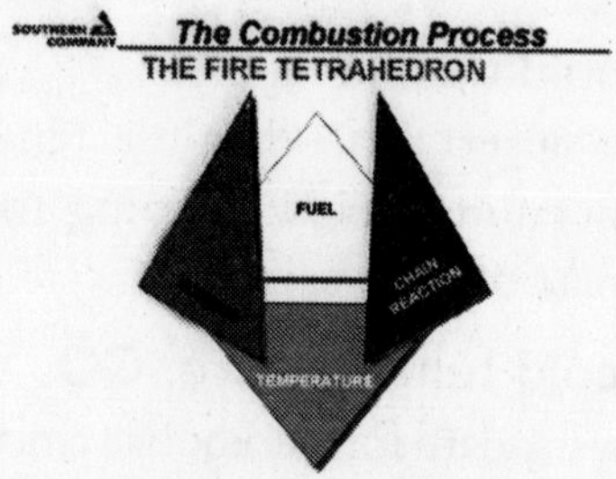

■ **combustion chamber**

the actual compartment where waste is burned in an incinerator.

■ **combustion product**

substance produced during the burning or oxidation of a material.

■ **command post**

facility located at a safe distance upwind from an accident site, where the on-scene coordinator, responders, and technical representatives make response decisions, deploy manpower and equipment, maintain liaison with news media, and handle communications.

■ **command-and-control regulations**

specific requirements prescribing how to comply with specific standards defining acceptable levels of pollution.

commensalism
a relationship in which a guest organism lives on or in a host organism. The guest organism benefits from the relationship, but the host is neither helped nor harmed.

commercial waste
all solid waste emanating from business establishments such as stores, markets, office buildings, restaurants, shopping centers, and theaters.

commercial waste management facility
a treatment, storage, disposal, or transfer facility which accepts waste from a variety of sources, as compared to a private facility which normally manages a limited waste stream generated by its own operations.

commingled recyclables
mixed recyclables that are collected together.

comminuter
a machine that shreds or pulverises solids to make waste treatment easier.

comminution
mechanical shredding or pulverising of waste. Used in both solid waste management and wastewater treatment.

common sense initiative
voluntary program to simplify environmental regulation to achieve cleaner, cheaper, smarter results, starting with six major industry sectors.

community
an integrated group of species inhabiting a given area; the organisms within a community influence one another's distribution, abundance, and evolution. (A Human Community is a social group of any size whose members reside in a specific locality.).

community water system
a public water system which serves at least 15 service connections used by year-round residents or regularly serves at least 25 year-round residents.

compact fluorescent lamp
small fluorescent lamps used as more efficient alternatives

to incandescent lighting. Also called PL, CFL, Twin-Tube, or BIAX lamps.

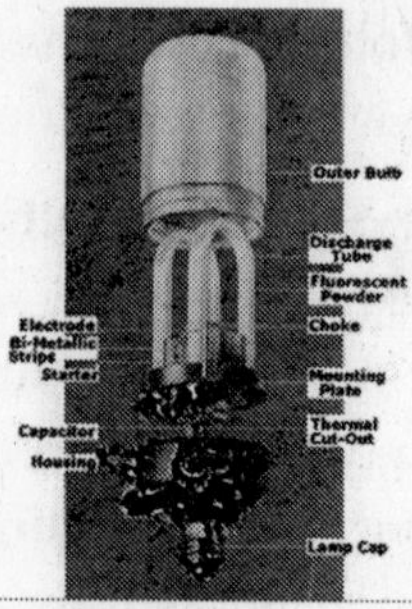

■ compaction
reduction of the bulk of solid waste by rolling and tamping.

■ comparative advantage
relative superiority with which a region or state may produce a good or service.

■ comparative risk assessment
process that generally uses the judgement of experts to predict effects and set priorities among a wide range of environmental problems.

■ complete treatment
a method of treating water that consists of the addition of coagulant chemicals, flash mixing, coagulation-flocculation, sedimentation, and filtration. Also called conventional filtration.

■ compliance coal
any coal that emits less than 1.2 pounds of sulphur dioxide per million Btu when burned. Also known as low sulphur coal.

■ compliance coating
a coating whose volatile organic compound content does not exceed that allowed by regulation.

■ compliance monitoring
collection and evaluation of data, including self-monitoring reports, and verification to show whether pollutant concentrations and loads contained in permitted discharges are in compliance with the limits and conditions specified in the permit.

■ compliance schedule
a negotiated agreement between a pollution source and a government agency that specifies dates and procedures by which a source will reduce emissions and, thereby, comply with a regulation.

■ composite sample
a series of water samples taken over a given period of

time and weighted by flow rate.

■ **compost**

the relatively stable humus material that is produced from a composting process in which bacteria in soil mixed with garbage and degradable trash break down the mixture into organic fertiliser.

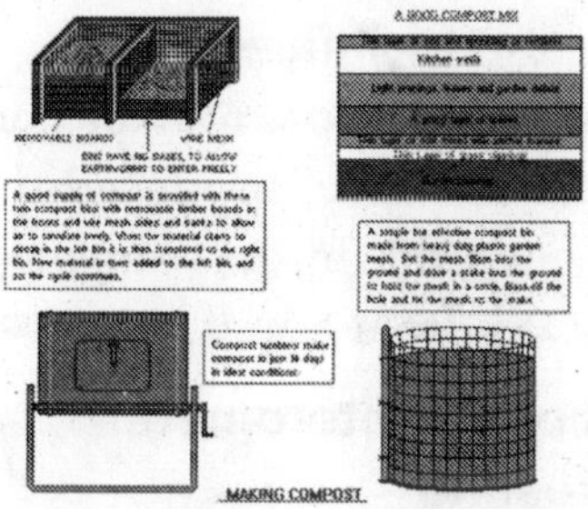

■ **composting**

the controlled biological decomposition of organic material in the presence of air to form a humus-like material. Controlled methods of composting include mechanical mixing and aerating, ventilating the materials by dropping them through a vertical series of aerated chambers, or placing the compost in piles out in the open air and mixing it or turning it periodically.

■ **composting facilities**

1. an offsite facility where the organic component of municipal solid waste is decomposed under controlled conditions; 2.an aerobic process in which organic materials are ground or shredded and then decomposed to humus in windrow piles or in mechanical digesters, drums, or similar enclosures.

■ **Compressed Natural Gas (CNG)**

an alternative fuel for motor vehicles; considered one of the cleanest because of low hydrocarbon emissions and its vapours are relatively non-ozone producing. However, vehicles fuelled with CNG do emit a significant quantity of nitrogen oxides.

■ **concentration**

the relative amount of a sub stance mixed with another substance. An example is five ppm of carbon monoxide in air or 1 mg/l of iron in water.

■ **condensate**

1. liquid formed when warm landfill gas cools as it travels through a collection system.

2. water created by cooling steam or water vapour.

■ **condensate return system**
system that returns the heated water condensing within steam piping to the boiler and thus saves energy.

■ **condensation**
the process by which a gas, such as water vapour, changes to a liquid when cooled; also the water that results from this process.

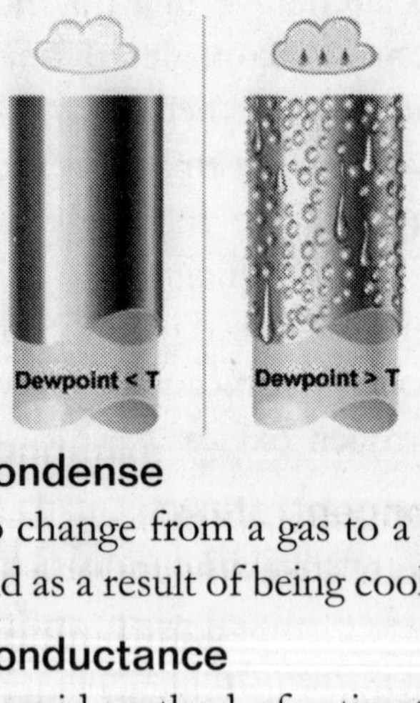

■ **condense**
to change from a gas to a liquid as a result of being cooled.

■ **conductance**
a rapid method of estimating the dissolved solids content of water supply by determining the capacity of a water sample to carry an electrical current. Conductivity is a measure of the ability of a solution to carry and electrical current.

■ **conductivity**
a measure of the ability of a solution to carry an electrical current.

■ **cone of depression**
a depression in the water table that develops around a pumped well.

■ **cone of influence**
the depression, roughly conical in shape, produced in a water table by the pumping of water from a well.

■ **cone penterometer testing**
a direct push system used to measure lithology based on soil penetration resistance. Sensors in the tip of the cone of the DP rod measure tip resistance and side-wall friction, transmitting electrical signals to digital processing equipment on the ground surface.

■ **confidential statement of formula**
a list of the ingredients in a new pesticide or chemical formulation. The list is submitted at the time for application

for registration or change in formulation.

■ confined aquifer

an aquifer in which ground water is confined under pressure which is significantly greater than atmospheric pressure.

■ confluent growth

a continuous bacterial growth covering all or part of the filtration area of a membrane filter in which the bacteria colonies are not discrete.

■ coniferous forest

a forest containing coniferous plants and existing where winters are very cold, summers are brief, and rainfall is low, such as in the northern parts of North America, Europe and Asia, and in the world's mountain regions. Also called a boreal forest, northern coniferous forest, or taiga.

■ coniferous plant

a plant, usually an evergreen, whose seeds are produced in cones and that typically has needles instead of leaves.

■ conservation

the management of human use of the biosphere so that it may yield the greatest sustainable benefit to current generations while maintaining its potential to meet the needs and aspirations of future generations: Thus conservation is positive, embracing preservation, maintenance, sustainable utilisation, restoration, *and* enhancement of the natural environment.

■ conservation biology

multidisciplinary science created to deal with the crisis of maintaining the genes, species, communities, and ecosystems that make up earth's biological diversity. Its goals are to investigate human impacts on biodiversity and to develop practical approaches to preserving biodiversity and ecological integrity.

■ conservation easement

easement restricting a landowner to land uses that that

are compatible with long-term conservation and environmental values.

■ **conservation of biodiversity**
the management of human interactions with genes, species, and ecosystems so as to provide the maximum benefit to the present generation while maintaining their potential to meet the needs and aspirations of future generations; encompasses elements of saving, studying, and using biodiversity.

■ **conservation-tillage farming**
crop cultivation in which the soil is disturbed little (minimum-tillage farming) or not at all (no-till farming) to reduce soil erosion, lower labor costs, and save energy.

■ **conserve**
to protect from waste or destruction.

■ **constituent(s) of concern**
specific chemicals that are identified for evaluation in the site assessment process.

■ **construction and demolition waste**
waste building materials, dredging materials, tree stumps, and rubble resulting from construction, remodeling, repair, and demolition of homes, commercial buildings and other structures and pavements. May contain lead, asbestos, or other hazardous substances.

■ **consumers**
organisms (specifically, animals) that are not able to produce their own food and must eat other organisms.

■ **consumptive water use**
water removed from available supplies without return to a water resources system, e.g. water used in manufacturing, agriculture, and food preparation.

■ **contact pesticide**
a chemical that kills pests when it touches them, instead of by ingestion. Also, soil that contains the minute skeletons of certain algae that scratch and dehydrate waxy-coated insects.

■ **contaminant**
any physical, chemical, biological, or radiological substance or matter that has an

adverse effect on air, water, or soil.

■ **contamination**
introduction into water, air, and soil of microorganisms, chemicals, toxic substances, wastes, or wastewater in a concentration that makes the medium unfit for its next intended use. Also applies to surfaces of objects, buildings, and various household and agricultural use products.

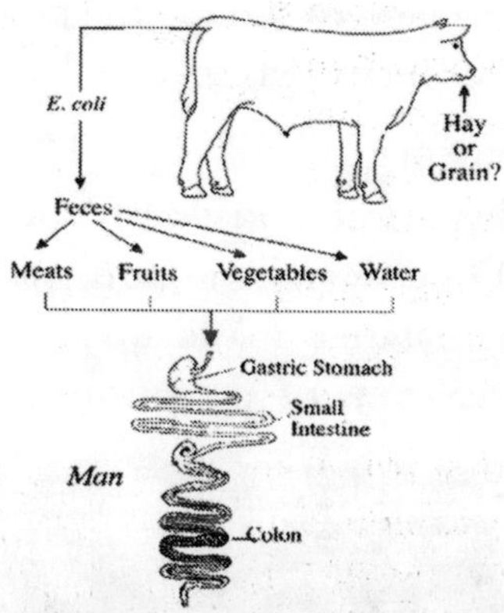

■ **continuous discharge**
a routine release to the environment that occurs without interruption, except for infrequent shutdowns for maintenance, process changes, etc.

■ **continuous sample**
a flow of water, waste or other material from a particular place in a plant to the location where samples are collected for testing. May be used to obtain grab or composite samples.

■ **contour farming**
the practice of planting crops across a hill rather than up and down the hill to prevent soil erosion by water.

■ **contour ploughing**
soil tilling method that follows the shape of the land to discourage erosion.

■ **contour strip farming**
a kind of contour farming in which row crops are planted in strips, between alternating strips of close-growing, erosion-resistant forage crops.

■ **controlled reaction**
a chemical reaction under temperature and pressure conditions maintained within safe

limits to produce a desired product or process.

■ conventional pollutants
statutorily listed pollutants understood well by scientists. These may be in the form of organic waste, sediment, acid, bacteria, viruses, nutrients, oil and grease, or heat.

■ conventional site assessment
assessment in which most of the sample analysis and interpretation of data is completed off-site; process usually requires repeated mobilisation of equipment and staff in order to fully determine the extent of contamination.

■ conventional systems
systems that have been traditionally used to collect municipal wastewater in gravity sewers and convey it to a central primary or secondary treatment plant prior to discharge to surface waters.

■ conventional tilling
tillage operations considered standard for a specific location and crop and that tend to bury the crop residues; usually considered as a base for determining the cost effectiveness of control practices.

■ conveyance loss
water loss in pipes, channels, conduits, ditches by leakage or evaporation.

■ cooling electricity use
amount of electricity used to meet the building cooling load.

■ cooling tower
a structure that helps remove heat from water used as a coolant; e.g., in electric power generating plants.

■ coral
live coral is made of polyps. Dead coral is a hard, stony substance made up of the skeletons of polyps.

■ coral bleaching
the loss of colour from a coral as it expels its zooxanthellae-usually a stress response.

■ core

1.the uranium-containing heart of a nuclear reactor, where energy is released. 2.the innermost part of the earth, center.

■ corrosion

the dissolution and wearing away of metal caused by a chemical reaction such as between water and the pipes, chemicals touching a metal surface, or contact between two metals.

■ corrosive

a chemical agent that reacts with the surface of a material causing it to deteriorate or wear away.

■ cosmopolitan

widely distributed over the globe.

■ cost recovery

a legal process by which potentially responsible parties who contributed to contamination at a Superfund site can be required to reimburse the Trust Fund for money spent during any cleanup actions by the federal government.

■ cost sharing

a publicly financed program through which society, as a beneficiary of environmental protection, shares part of the cost of pollution control with those who must actually install the controls. In Superfund, for example, the government may pay part of the cost of a cleanup action with those responsible for the pollution paying the major share.

■ cost-benefit analysis

estimates and comparison of short-term and long-term costs (losses) and benefits (gains) from an economic decision. If the estimated benefits exceed the estimated costs, the decision to buy an economic good or provide a public good is considered worthwhile.

■ cover crop

a crop that provides temporary protection for delicate seedlings and/or provides a cover canopy for seasonal soil protection and improvement between normal crop production periods.

■ cover material
soil used to cover compacted solid waste in a sanitary landfill.

■ cradle-to-grave or manifest system
a procedure in which hazardous materials are identified and followed as they are produced, treated, transported, and disposed of by a series of permanent, linkable, descriptive documents (e.g. manifests). Commonly referred to as the cradle-to-grave system.

■ critical effect
the first adverse effect, or its known precursor, that occurs as a dose rate increases. Designation is based on evaluation of overall database.

■ crop consumptive use
the amount of water transpired during plant growth plus what evaporated from the soil surface and foliage in the crop area.

■ crop rotation
the practice of planting different crops each season so that the soil has time to renew itself.

■ cross contamination
the movement of underground contaminants from one level or area to another due to invasive subsurface activities.

■ cross-connection
any actual or potential connection between a drinking water system and an unapproved water supply or other source of contamination.

■ crumb rubber
ground rubber fragments the size of sand or silt used in rubber or plastic products, or processed further into reclaimed rubber or asphalt products.

■ cryogenics
the branch of physics relating to the effects and production of very low temperatures; as applied to living organisms, preservation in a dormant state by freezing, drying, or both.

■ cryptosporidium
a protozoan microbe associated with the disease cryptosporidiosis in man. The disease can be transmitted through ingestion of drinking water, person-to-person con-

tact, or other pathways, and can cause acute diarrhoea, abdominal pain, vomiting, fever, and can be fatal as it was in the Milwaukee episode.

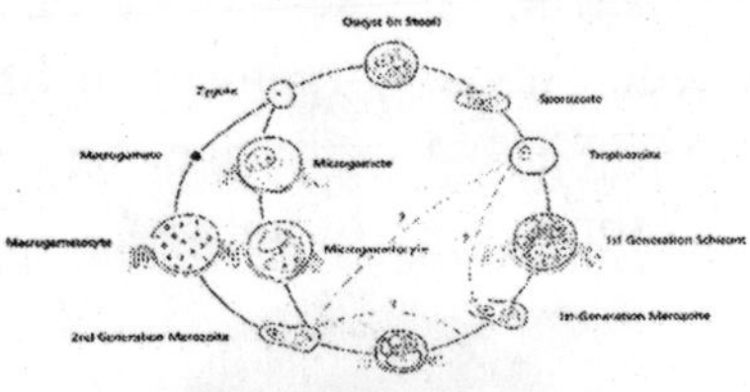

Cryptosporidium parvum Life cycle

■ **cubic feet per minute**
a measure of the volume of a substance flowing through air within a fixed period of time. With regard to indoor air, refers to the amount of air, in cubic feet, that is exchanged with outdoor air in a minute's time; i.e. the air exchange rate.

■ **cullet**
crushed glass.

■ **cultivar**
a cultivated variety (genetic strain) of a domesticated crop plant.

■ **cultural diversity**
variety or multiformity of human social structures, belief systems, and strategies for adapting to situations in different parts of the world.

■ **cultural eutrophication**
increasing rate at which water bodies "die" by pollution from human activities.

■ **cultures and stocks**
infectious agents and associated biologicals including cultures from medical and pathological laboratories; cultures and stocks of infectious agents from research and industrial laboratories; waste from the production of biologicals; discarded live and attenuated vaccines; and culture dishes and devices used to transfer, inoculate, and mix cultures.

■ **cumulative ecological risk assessment**
consideration of the total ecological risk from multiple stressors to a given eco-zone.

■ **cumulative exposure**
the sum of exposures of an organism to a pollutant over a period of time.

■ **cumulative working level months**
the sum of lifetime exposure to radon working levels expressed in total working level months.

■ **cumulonimbus**
rain clouds which can result in breif or less extended rainfall and thunderstorms.

■ **cumulus**
puffy clouds which mark unstable air and can be found at any level in the atmosphere.

■ **curb stop**
a water service shutoff valve located in a water service pipe near the curb and between the water main and the building.

■ **curbside collection**
method of collecting recyclable materials at homes, community districts or businesses.

■ **cutie-pie**
an instrument used to measure radiation levels.

■ **cuttings**
spoils left by conventional drilling with hollow stem auger or rotary drilling equipment.

■ **cyclone collector**
a device that uses centrifugal force to remove large particles from polluted air.

■ **cyst**
a small, capsule-like sac that encloses an organism in its resting or larval stage, e.g., a resting spore of an alga.

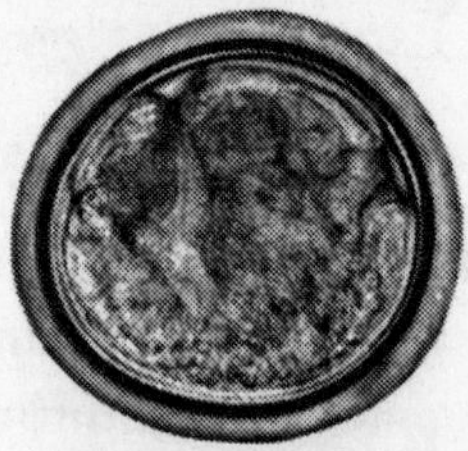

Larval cyst (Microphallus)

■ **data quality objectives**
qualitative and quantitative statements of the overall level of uncertainty that a decision-maker will accept in results or decisions based on environmental data. They provide the statistical framework for planning and managing environmental data operations consistent with user's needs.

■ **day tank**
another name for de-aerating tank. (See age tank.).

■ **DDT**
the first chlorinated hydrocarbon insecticide chemical

name Dichloro-Diphenyl-Trichloroethane. It has a half-life of 15 years and can collect in fatty tissues of certain animals.

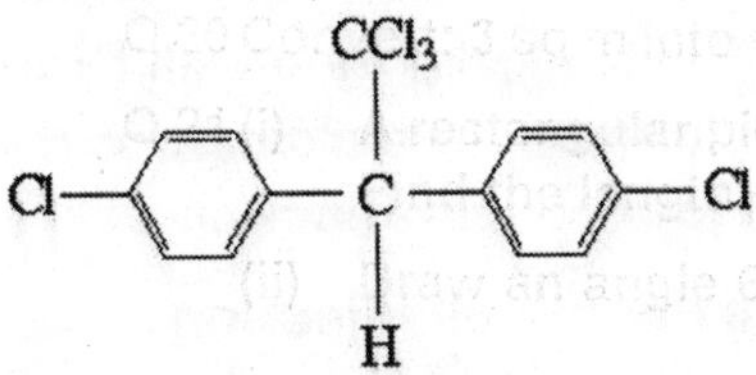

■ **dead end**
the end of a water main which is not connected to other parts of the distribution system.

■ **deadmen**
anchors drilled or cemented into the ground to provide additional reactive mass for DP sampling rigs.

■ **debt-for-nature swap**
agreement in which a certain amount of foreign debt is cancelled in exchange for local currency investments that will improve natural resource management or protect certain areas in the debtor country from harmful development.

■ **decant**
to draw off the upper layer of liquid after the heaviest material (a solid or another liquid) has settled.

■ **decay products**
degraded radioactive materials, often referred to as "daughters" or "progeny"; radon decay products of most concern from a public health standpoint are polonium-214 and polonium-218.

■ **dechlorination**
removal of chlorine from a substance.

■ **deciduous forest**
a forest containing deciduous plants and existing where temperatures are mild and rainfall is abundant.

■ **deciduous plant**
a plant that sheds all or nearly all its leaves each year.

decompose
to rot or decay as a result of being broken down by micro-organisms.

decomposers
organisms such as bacteria and fungi that decompose dead plants and animals.

decomposition
the breakdown of organic materials by organisms in the environment, releasing energy and simple organic and inorganic compounds. About 10 percent of the energy that enters living systems through photosynthesis in plants passes to herbivores, and a fraction of this energy then passes to carnivores. Whether feeding on living or non-living material, however, the *detritivores* (the organisms consuming non-living material, such as many fungi, bacteria, and earthworms) and consumers break down organic material (such as sugars and proteins) to obtain energy for their own growth, thereby returning the inorganic components (the nutrients) to the environment, where they are again available to plants.

decontamination
removal of harmful substances such as noxious chemicals, harmful bacteria or other organisms, or radioactive material from exposed individuals, rooms and furnishings in buildings, or the exterior environment.

deep-well injection
deposition of raw or treated, filtered hazardous waste by pumping it into deep wells, where it is contained in the pores of permeable subsurface rock.

deflocculating agent
a material added to a suspension to prevent settling.

defluoridation
the removal of excess fluoride in drinking water to prevent the staining of teeth.

defoliant
an herbicide that removes leaves from trees and growing plants.

deforestation
removal of trees from a forested area without adequate replanting.

■ degasification

a water treatment that removes dissolved gases from the water.

■ degree-day

a rough measure used to estimate the amount of heating required in a given area; is defined as the difference between the mean daily temperature and 65 degrees Fahrenheit. Degree-days are also calculated to estimate cooling requirements.

■ demineralisation

a treatment process that removes dissolved minerals from water.

■ demographic transition

hypothesis that countries, as they become industrialized, have declines in death rates followed by declines in birth rates.

■ demography

the rate of growth and the age structure of populations, and the processes that determine these properties.

■ denitrification

the biological reduction of nitrate to nitrogen gas by denitrifying bacteria in soil.

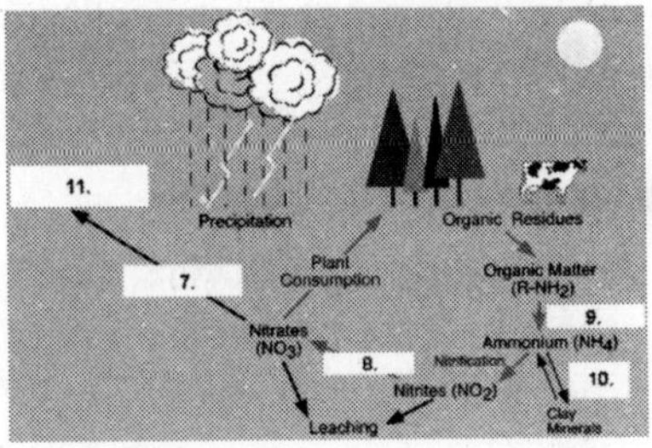

■ dense non-aqueous phase liquid

non-aqueous phase liquids such as chlorinated hydrocarbon solvents or petroleum fractions with a specific gravity greater than 1.0 that sink through the water column until they reach a confining layer. Because they are at the bottom of aquifers instead of floating on the water table, typical monitoring wells do not indicate their presence.

■ density

a measure of how heavy a specific volume of a solid, liquid, or gas is in comparison to water depending on the chemical.

■ depletion curve

in hydraulics, a graphical representation of water depletion from storage-stream channels, surface soil, and groundwater. A depletion curve can be drawn for base flow, direct runoff, or total flow.

■ depressurisation

a condition that occurs when the air pressure inside a structure is lower that the air pressure outdoors. Depressurisation can occur when household appliances such as fireplaces or furnaces, that consume or exhaust house air, are not supplied with enough makeup air. Radon may be drawn into a house more rapidly under depressurised conditions.

■ dermal absorption/ penetration

process by which a chemical penetrates the skin and enters the body as an internal dose.

■ dermal exposure

contact between a chemical and the skin.

■ dermal toxicity

the ability of a pesticide or toxic chemical to poison people or animals by contact with the skin.

■ DES

a synthetic estrogen, diethylstilbestrol is used as a growth stimulant in food animals. Residues in meat are thought to be carcinogenic.

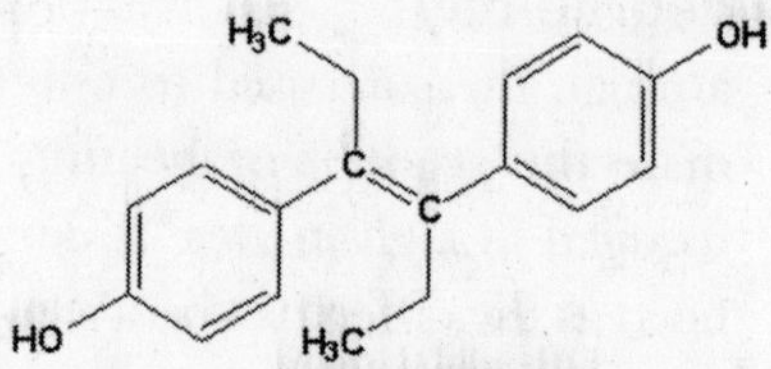

Diethylstilbestrol

■ desalination

(1) removing salts from ocean or brackish water by using various technologies. (2) Removal of salts from soil by artificial means, usually leaching.

■ desert

a land area that receives less than 10 inches (25 cm) of precipitation a year, that loses more water through evaporation than it gains from precipitation, and that has high summer temperatures.

■ desertification

conversion of rangeland, rainfed cropland to desert-like land, with a drop in agricultural productivity of 10% or more. It is usually caused by a combination of overgrasing, soil erosion, prolonged drought, and climate change.

■ desiccant

a chemical agent that absorbs moisture; some desiccants are capable of drying out plants or insects, causing death.

■ design capacity

the average daily flow that a treatment plant or other facility is designed to accommodate.

■ designated pollutant

an air pollutant which is neither a criteria nor hazardous pollutant, as described in the Clean Air Act, but for which new source performance standards exist. The Clean Air Act does require states to control these pollutants, which include acid mist, total reduced sulphur (TRS), and fluorides.

■ designated uses

those water uses identified in state water quality standards that must be achieved and maintained as required under the Clean Water Act. Uses can include cold water fisheries, public water supply, and irrigation.

■ designer bugs

popular term for microbes developed through biotechnology that can degrade specific toxic chemicals at their source in toxic waste dumps or in ground water.

■ destination facility

the facility to which regulated medical waste is shipped for treatment and destruction, incineration, and/or disposal.

■ destratification

vertical mixing within a lake or reservoir to totally or partially eliminate separate layers of temperature, plant, or animal life.

■ destroyed medical waste

regulated medical waste that has been ruined, torn apart, or mutilated through thermal treatment, melting, shredding, grinding, tearing, or breaking, so that it is no longer generally recognised as medical waste, but has not yet been treated (excludes compacted regulated medical waste).

■ destruction and removal

efficiency
a percentage that represents the number of molecules of a compound removed or destroyed in an incinerator relative to the number of molecules entering the system (e.g. a DRE of 99.99 percent means that 9,999 molecules are destroyed for every 10,000 that enter; 99.99 percent is known as "four nines."

■ **destruction facility**
a facility that destroys regulated medical waste.

■ **desulphurisation**
removal of sulphur from fossil fuels to reduce pollution.

■ **detectable leak rate**
the smallest leak (from a storage tank), expressed in terms of gallons- or liters-per-hour, that a test can reliably discern with a certain probability of detection or false alarm.

■ **detection criterion**
a predetermined rule to ascertain whether a tank is leaking or not. Most volumetric tests use a threshold value as the detection criterion. (See volumetric tank tests.).

■ **detection limit**
the lowest concentration of a chemical that can reliably be distinguished from a zero concentration.

■ **detention time**
1. The theoretical calculated time required for a small amount of water to pass through a tank at a given rate of flow. 2. The actual time that a small amount of water is in a settling basin, flocculating basin, or rapid-mix chamber. 3. In storage reservoirs, the length of time water will be held before being used.

■ **detergent**
synthetic washing agent that helps to remove dirt and oil. Some contain compounds which kill useful bacteria and encourage algae growth when they are in wastewater that reaches receiving waters.

■ **detritus**
accumulated organic debris from dead organisms, often an important source of nutrients in a food web.

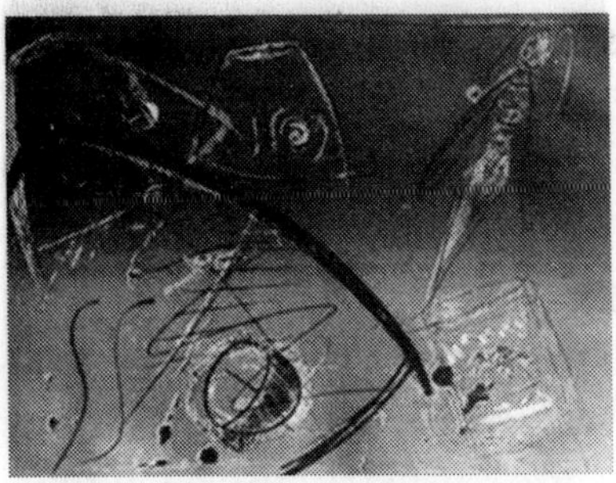

■ **detrivore**

any organism which obtains most of its nutrients from the detritus in an ecosystem.

■ **development effects**

adverse effects such as altered growth, structural abnormality, functional deficiency, or death observed in a developing organism.

■ **dewater**

1.remove or separate a portion of the water in a sludge or slurry to dry the sludge so it can be handled and disposed of. 2. Remove or drain the water from a tank or trench.

■ **dewpoint**

the temperature at which gaseous water condenses into visible water vapour, fog or clouds.

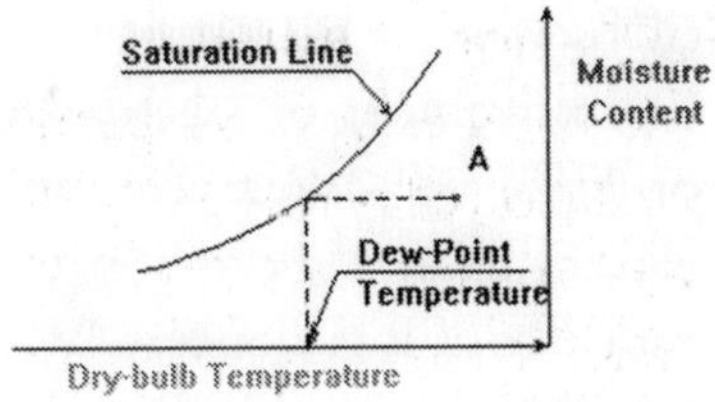

■ **diameter**

the distance from one end to another thru the center; as the diameter of the earth

■ **diatomaceous earth (diatomite)**

a chalk-like material (fossilised diatoms) used to filter out solid waste in wastewater treatment plants; also used as an active ingredient in some powdered pesticides.

■ **diazinon**

an insecticide.

■ **dibenzofurans**

a group of organic compounds, some of which are toxic.

■ **dicofol**

a pesticide used on citrus fruits.

■ **diffused air**

a type of aeration that forces oxygen into sewage by pumping air through perforated pipes inside a holding tank.

■ diffusion
the movement of suspended or dissolved particles (or molecules) from a more concentrated to a less concentrated area. The process tends to distribute the particles or molecules more uniformly.

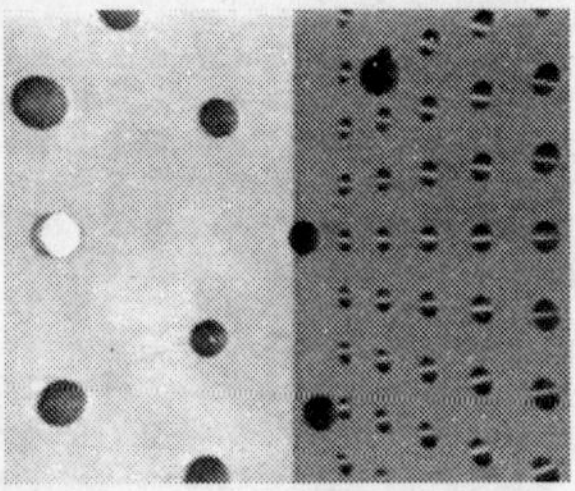

■ digester
in wastewater treatment, a closed tank; in solid-waste conversion, a unit in which bacterial action is induced and accelerated in order to break down organic matter and establish the proper carbon to nitrogen ratio.

■ digestion
the biochemical decomposition of organic matter, resulting in partial gasification, liquefaction, and mineralisation of pollutants.

■ dike
a low wall that can act as a barrier to prevent a spill from spreading.

■ diluent
any liquid or solid material used to dilute or carry an active ingredient.

■ dilute
to lessen the strength of a material by mixing it with another material, usually water.

■ dilution ratio
the relationship between the volume of water in a stream and the volume of incoming water. It affects the ability of the stream to assimilate waste.

■ dimictic
lakes and reservoirs that freeze over and normally go through two stratifications and two mixing cycles a year.

■ dinoseb
a herbicide that is also used as a fungicide and insecticide.

■ dioxin
a synthetic, organic chemical of the chlorinated hydrocarbon class. It is one of the most toxic compounds known to humans, having many harmful effects, including induction of cancer and birth defects, even in extremely minute concentrations. It has

become a widespread environmental pollutant because of the use of certain herbicides that contain dioxin as a contaminant.

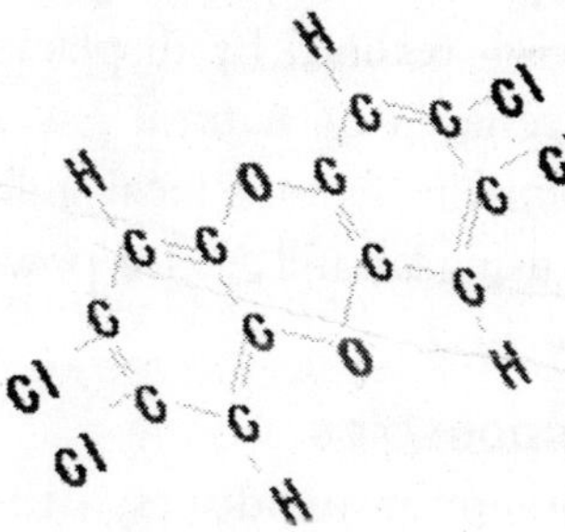

■ **direct discharger**

a municipal or industrial facility which introduces pollution through a defined conveyance or system such as outlet pipes; a point source.

■ **direct filtration**

a method of treating water which consists of the addition of coagulant chemicals, flash mixing, coagulation, minimal flocculation, and filtration. Sedimentation is not uses.

■ **direct push**

technology used for performing subsurface investigations by driving, pushing, and/or vibrating small-diameter hollow steel rods into the ground/ Also known as direct drive, drive point, or push technology.

■ **direct runoff**

water that flows over the ground surface or through the ground directly into streams, rivers, and lakes.

■ **dirty fallout**

air pollutants dropped by prevailing winds.

■ **discharge**

flow of surface water in a stream or canal or the outflow of ground water from a flowing artesian well, ditch, or spring. Can also apply tp discharge of liquid effluent from a facility or to chemical emissions into the air through designated venting mechanisms.

■ **disease**

organisms suffer from disease when their normal function is impaired by some genetic disorder, or more often from the activity of a parasite or other organism living within them. Many diseases are caused by viruses, bacteria, or fungi.

■ **disinfectant**

a chemical or physical process that kills pathogenic organisms in water, air, or on surfaces. Chlorine is often used to disinfect sewage treatment effluent, water supplies, wells, and swimming pools.

■ **disinfectant by-product**
a compound formed by the reaction of a disinfenctant such as chlorine with organic material in the water supply; a chemical byproduct of the disinfection process..

■ **disinfectant time**
the time it takes water to move from the point of disinfectant application (or the previous point of residual disinfectant measurement) to a point before or at the point where the residual disinfectant is measured. In pipelines, the time is calculated by dividing the internal volume of the pipe by he maximum hourly flow rate; within mixing basins and storage reservoirs it is determined by tracer studies of an equivalent demonstration.

■ **dispersant**
a chemical agent used to break up concentrations of organic material such as spilled oil.

■ **disperse**
to spread to another location.

■ **displacement savings**
saving realised by displacing purchases of natural gas or electricity from a local utility by using landfill gas for power and heat.

■ **disposables**
consumer products, other items, and packaging used once or a few times and discarded.

■ **disposal**
final placement or destruction of toxic, radioactive, or other wastes; surplus or banned pesticides or other chemicals; polluted soils; and drums containing hazardous materials from removal actions or accidental releases. Disposal may be accomplished through use of approved secure landfills, surface impoundments, land farming, deep-well injection, ocean dumping, or incineration.

■ **disposal facilities**
repositories for solid waste, including landfills and com-

bustors intended for permanent containment or destruction of waste materials. Excludes transfer stations and composting facilities.

■ dissolved oxygen (DO)
the oxygen freely available in water, vital to fish and other aquatic life and for the prevention of odours. DO levels are considered a most important indicator of a water body's ability to support desirable aquatic life. Secondary and advanced waste treatment is generally designed to ensure adequate DO in waste-receiving waters.

■ dissolved solids
disintegrated organic and inorganic material in water. Excessive amounts make water unfit to drink or use in industrial processes.

■ distance
the measurement from one point to another.

■ distillation
the act of purifying liquids through boiling, so that the steam or gaseous vapours condense to a pure liquid. Pollutants and contaminants may remain in a concentrated residue.

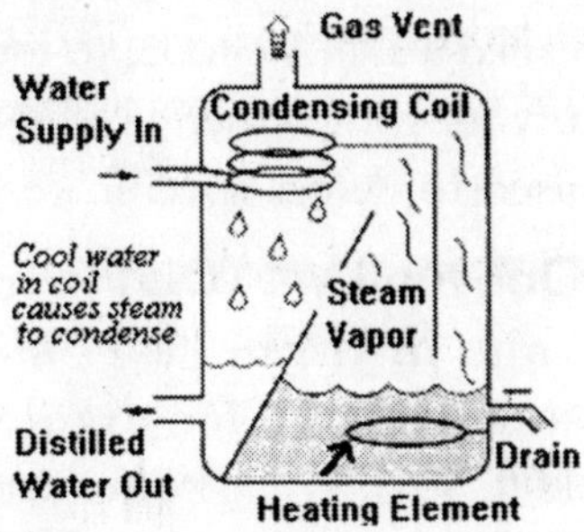

■ disturbance
any event or series of events that disrupt ecosystem, community, or population structure and alters the physical environment.

■ diversion
1. Use of part of a stream flow as water supply. 2. A channel with a supporting ridge on the lower side constructed across a slope to divert water at a non-erosive velocity to sites where it can be used and disposed of.

■ diversion rate
the percentage of waste materials diverted from traditional disposal such as land filling or incineration to be recycled, composted, or re-used.

■ DNA hybridisation
use of a segment of DNA, called a DNA probe, to identify its complementary DNA; used to detect specific genes.

■ Dobson Unit (DU)
units of ozone level measurement. If, for example, 100 DU of ozone were brought to the earth's surface they would form a layer one millimetre thick. Ozone levels vary geographically, even in the absence of ozone depletion.

■ domestic application
pesticide application in and around houses, office buildings, motels, and other living or working areas.

■ donor control
a predator-prey interaction in which the predator does not control the prey population size.

■ dormant
inactive.

■ dosage/dose
1.the actual quantity of a chemical administered to an organism or to which it is exposed. 2. The amount of a substance that reaches a specific tissue (e.g. the liver). 3. the amount of a substance available for interaction with metabolic processes after crossing the outer boundary of an organism.

■ dose equivalent
the product of the absorbed dose from Ionising radiation and such factors as account for biological differences due to the type of radiation and its distribution in the body in the body.

■ dose rate
in exposure assessment, dose per time unit (e.g. mg/day), sometimes also called dosage.

■ dose response
shifts in toxicological responses of an individual (such as alterations in severity) or populations (such as alterations in incidence) that are related to changes in the dose of any given substance.

■ dose response curve
graphical representation of the relationship between the dose of a stressor and the biological response thereto.

■ **dose-response assessment**

1. Estimating the potency of a chemical.

2. In exposure assessment, the process of determining the relationship between the dose of a stressor and a specific biological response.

3. Evaluating the quantitative relationship between dose and toxicological responses.

■ **dose-response relationship**

the quantitative relationship between the amount of exposure to a substance and the extent of toxic injury or disease produced.

■ **dosimeter**

an instrument to measure dosage; many so-called dosimeters actually measure exposure rather than dosage. Dosimetry is the process or technology of measuring and/or estimating dosage.

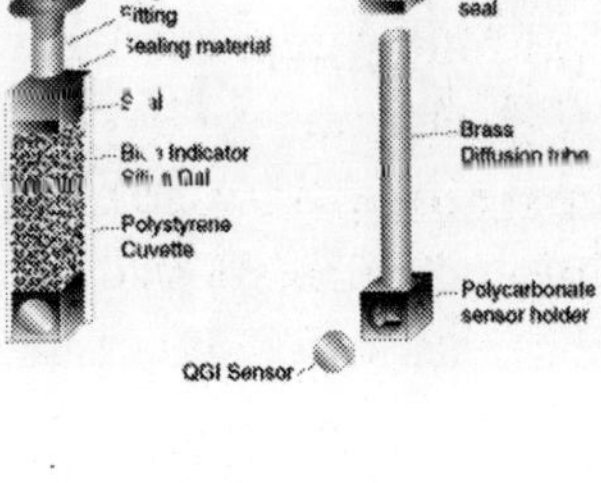

■ **downgradient**

the direction that groundwater flows; similar to "downstream" for surface water.

■ **downstream processors**

industries dependent on crop production (e.g. canneries and food processors).

■ **DP hole**

hole in the ground made with DP equipment. (See direct push.).

■ **draft**

1. the act of drawing or removing water from a tank or reservoir. 2. the water which is drawn or removed.

■ **drainage**

improving the productivity of agricultural land by removing excess water from the soil by such means as ditches or subsurface drainage tiles.

■ **drainage basin**

the area of land that drains water, sediment, and dissolved materials to a common outlet at some point along a stream channel.

■ **drainage well**

a well drilled to carry excess water off agricultural fields.

Because they act as a funnel from the surface to the groundwater below. Drainage wells can contribute to groundwater pollution.

■ **drawdown**

1. The drop in the water table or level of water in the ground when water is being pumped from a well. 2. The amount of water used from a tank or reservoir. 3. The drop in the water level of a tank or reservoir.

■ **dredging**

removal of mud from the bottom of water bodies. This can disturb the ecosystem and causes silting that kills aquatic life. Dredging of contaminated muds can expose biota to heavy metals and other toxics.

■ **drilling fluid**

fluid used to lubricate the bit and convey drill cuttings to the surface with rotary drilling equipment. Usually composed of bentonite slurry or muddy water. Can become contaminated, leading to cross contamination, and may require special disposal. Not used with DP methods.

■ **drinking water equivalent level**

protective level of exposure related to potentially non-carcinogenic effects of chemicals that are also known to cause cancer.

■ **drinking water state revolving fund**

the Fund provides capitalisation grants to states to develop drinking water revolving loan funds to help finance system infrastructure improvements, assure source-water protection, enhance operation and management of drinking-water systems, and otherwise promote local water-system compliance and protection of public health.

■ **drip irrigation**

the practice of spraying water directly on the base of plants so that less water is needed to help them grow.

■ **drive casing**

heavy duty steel casing driven along with the sampling tool in cased DP systems. Keeps the hole open between sampling runs and is not removed until last sample has been collected.

■ **drive point profiler**

an exposed groundwater DP system used to collect multiple depth-discrete groundwater samples. Ports in the tip of the probe connect to an internal stainless steel or teflon tube that extends to the surface. Samples are collected via suction or air-lift methods. Deionised water is pumped down through the ports to prevent plugging while driving the tool to the next sampling depth.

■ **drop-off**

recyclable materials collection method in which individuals bring them to a designated collection site.

■ **drought**

an extended period of unusually low rainfall.

■ **dry deposits**

air pollutants that quickly fall to the ground without combining with moisture.

■ **dual-phase extraction**

active withdrawal of both liquid and gas phases from a well usually involving the use of a vacuum pump.

■ **dump**

a site used to dispose of solid waste without environmental controls.

■ **duplicate**

a second aliquot or sample that is treated the same as the original sample in order to determine the precision of the analytical method. (See aliquot.).

■ **dustfall jar**

an open container used to collect large particles from the air for measurement and analysis.

■ **dystrophic lakes**

acidic, shallow bodies of water that contain much humus and/or other organic matter; contain many plants but few fish.

■ earth
the planet where we live; see planet, and planetary facts

■ ecological community
the interaction of living organisms with their environment.

■ ecological efficiency
the percentage of energy in biomass produced by one trophic level that is incorporated into biomass by the next highest trophic level.

■ ecological entity
in ecological risk assessment, a general term referring to a species, a group of species, an ecosystem function or characteristic, or a specific habitat or biome.

■ ecological exposure
exposure of a non-human organism to a stressor.

■ ecological fitness
the number of a parent's young that live to reproduce; divided by two if sexual reproduction is involved.

■ ecological impact
the effect that a man-caused or natural activity has on living organisms and their nonliving (abiotic) environment.

■ ecological indicator
a characteristic of an ecosystem that is related to, or derived from, a measure of biotic or abiotic variable, that can provide quantitative information on ecological structure and function. An indicator can contribute to a measure of integrity and sustainability.

■ ecological integrity
a living system exhibits integrity if, when subjected to disturbance, it sustains and organises self-correcting ability to recover toward a biomass end-state that is normal for that system. End-states other than the pristine or naturally whole may be accepted as normal and good.

■ ecological risk assessment
the application of a formal framework, analytical process, or model to estimate the effects of human actions(s) on a natural resource and to interpret the significance of those effects in light of the uncertainties identified in each component of the assessment process. Such analy-

sis includes initial hazard identification, exposure and dose-response assessments, and risk characterisation.

■ **ecological succession**

process in which communities of plant and animal species in a particular area are replaced over time by a series of different and often more complex communities.

■ **ecological/environmental sustainability**

maintenance of ecosystem components and functions for future generations.

■ **ecologically sustainable development**

developments in which the total human population size and resource use in the world (or in a region) are limited to a level that does not exceed the carrying capacity or the existing natural capital and is therefore sustainable.

■ **ecologist**

a scientist who studies organisms and their environment.

■ **ecology**

the study of the relationships between organisms and their environments, including the interactions of living organisms with one another and with their non-living surroundings, the flow of matter and energy in an environment, and the structure and functions of nature. Also called bionomics . 2. The relationship between organisms and their environment. 3. The branch of sociology that is concerned with studying the relationships between human groups and their physical and social environments. Also called human ecology . 4. The study of the detrimental effects of modern civilisation on the environment, with a view toward prevention or reversal through conservation. A component of the field of human ecology .

■ **economic poisons**

chemicals used to control pests and to defoliate cash crops such as cotton.

■ **ecosphere**

the "bio-bubble" that contains life on earth, in surface waters, and in the air. (See biosphere.).

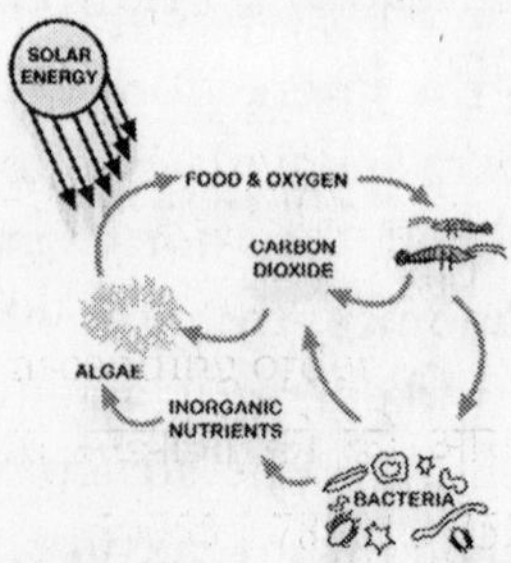

■ **ecosystem**
a distinct area that combines biotic communities and the abiotic environments with which they interact.

■ **ecosystem diversity**
variety of biomes and habitats occurring in the biosphere.

■ **ecosystem services**
services, vital to the support of human life, provided by intact natural ecosystems. These include the purification of air and water, detoxification and decomposition of wastes, regulation of climate, regeneration of soil fertility, and production and maintenance of biodiversity, from which key ingredients of our agricultural, pharmaceutical, and industrial enterprises are derived. Historically, the nature and value of Earth's life support systems have largely been ignored until their disruption or loss highlighted their importance. Read more about the value of ecosystem services and how they are measured.

■ **ecosystem structure**
attributes related to the instantaneous physical state of an ecosystem; examples include species population density, species richness or evenness, and standing crop biomass.

■ **ecotone**
the area where two or more ecosystems merge.

■ **ecotourism**
travel undertaken to witness sites or regions of unique natural or ecologic quality, or the provision of services to facilitate such travel.

■ **ecotype**
a genetically differentiated subpopulation that is restricted to a specific habitat.

■ **effluent**
wastewater—treated or untreated—that flows out of a treatment plant, sewer, or industrial outfall. Generally re-

fers to wastes discharged into surface waters.

■ effluent standard

see **effluent limitation.**

■ ejector

a device used to disperse a chemical solution into water being treated.

■ el nino southern oscillation (enso)

flip-flopping pressure systems in the South Pacific that trigger short-lived global changes in climate. Warm waters from the western Pacific move across the ocean, just below the equator, and significantly warm the eastern tropical Pacific.

■ electrodialysis

a process that uses electrical current applied to permeable membranes to remove minerals from water. Often used to desalinise salty or brackish water.

■ electromagnetic geophysical methods

ways to measure subsurface conductivity via low-frequency electromagnetic induction.

■ electrostatic precipitator (ESP)

a device that removes particles from a gas stream (smoke) after combustion occurs. The ESP imparts an electrical charge to the particles, causing them to adhere to metal plates inside the precipitator. Rapping on the plates causes the particles to fall into a hopper for disposal.

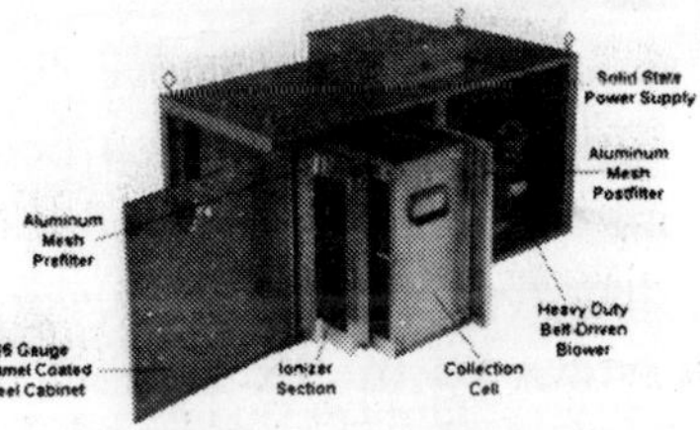

■ elevation

the height above sea level.

■ eMAP data

environmental monitoring data collected under the auspices of the Environmental Monitoring and Assessment Program. All EMAP data share the common attribute of being of known quality, having been collected in the context of explicit data quality objectives (DQOs) and a consistent quality assurance program.

■ **emergency (chemical)**
a situation created by an accidental release or spill of hazardous chemicals that poses a threat to the safety of workers, residents, the environment, or property.

■ **emergency episode**
see **air pollution episode.**

■ **emergency removal action**
1. steps take to remove contaminated materials that pose imminent threats to local residents (e.g. removal of leaking drums or the excavation of explosive waste.) 2. The state record of such removals.

■ **emergency response values**
concentrations of chemicals, published by various groups, defining acceptable levels for short-term exposures in emergencies.

■ **emergency suspension**
suspension of a pesticide product registration due to an imminent hazard. The action immediately halts distribution, sale, and sometimes actual use of the pesticide involved.

■ **emergent**
a tree with a canopy that forms about the general upper most continuous canopy.

■ **emergent layer**
a forest's upper layer, produced by the tallest trees.

■ **emerging disease**
the Institute for Medicine defines emerging and re-emerging diseases as "New, re-emerging, or drug-resistant infections whose incidence in humans has increased in the last two decades or whose incidence threatens to increase in the near future."

■ **emission**
pollution discharged into the atmosphere from smokestacks, other vents, and surface areas of commercial or industrial facilities; from residential chimneys; and from motor vehicle, locomotive, or aircraft exhausts.

■ **emission cap**
a limit designed to prevent projected growth in emissions

from existing and future stationary sources from eroding any mandated reductions. Generally, such provisions require that any emission growth from facilities under the restrictions be offset by equivalent reductions at other facilities under the same cap. (See emissions trading.).

■ emission factor

the relationship between the amount of pollution produced and the amount of raw material processed. For example, an emission factor for a blast furnace making iron would be the number of pounds of particulates per ton of raw materials.

■ emission inventory

a listing, by source, of the amount of air pollutants discharged into the atmosphere of a community; used to establish emission standards.

■ emission standard

the maximum amount of air polluting discharge legally allowed from a single source, mobile or stationary.

■ emissions trading

the creation of surplus emission reductions at certain stacks, vents or similar emissions sources and the use of this surplus to meet or redefine pollution requirements applicable to other emissions sources. This allows one source to increase emissions when another source reduces them, maintaining an overall constant emission level. Facilities that reduce emissions substantially may "bank" their "credits" or sell them to other facilities or industries.

■ emulsifier

a chemical that aids in suspending one liquid in another. Usually an organic chemical in an aqueous solution.

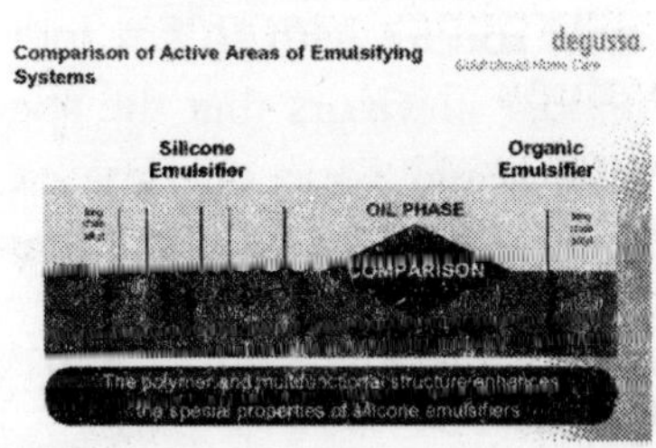

■ encapsulation

the treatment of asbestos-containing material with a liquid that covers the surface with a protective coating or embeds fibers in an adhesive matrix to prevent their release into the air.

■ enclosure

putting an airtight, impermeable, permanent barrier around asbestos-containing materials to prevent the release of asbestos fibers into the air.

■ end user

consumer of products for the purpose of recycling. Excludes products for re-use or combustion for energy recovery.

■ endangered

in immediate danger of becoming extinct.

■ endangered species

wild species with so few individual survivors that the species could soon become extinct in all or most of its natural range.

■ endemic

restricted to a specified region or locality.

■ end-of-the-pipe

technologies such as scrubbers on smokestacks and catalytic convertors on automobile tailpipes that reduce emissions of pollutants after they have formed.

■ endrin

a pesticide toxic to freshwater and marine aquatic life that produces adverse health effects in domestic water supplies.

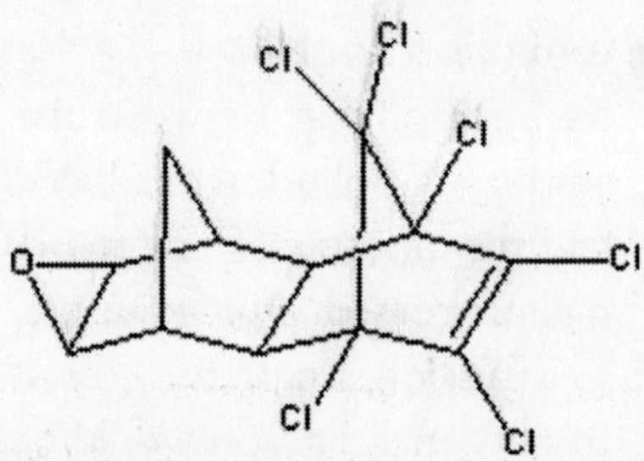

■ end-use product

a pesticide formulation for field or other end use. The label has instructions for use or application to control pests or regulate plant growth. The term excludes products used to formulate other pesticide products.

■ energy management system

a control system capable of monitoring environmental and system loads and adjusting HVAC operations accordingly in order to conserve energy while maintaining comfort.

energy recovery
obtaining energy from waste through a variety of processes (e.g. combustion).

engineered controls
method of managing environmental and health risks by placing a barrier between the contamination and the rest of the site, thus limiting exposure pathways.

enrichment
the addition of nutrients (e.g. nitrogen, phosphorus, carbon compounds) from sewage effluent or agricultural runoff to surface water, greatly increases the growth potential for algae and other aquatic plants.

entrain
to trap bubbles in water either mechanically through turbulence or chemically through a reaction.

environment
all external conditions and factors, living and nonliving (chemicals and energy), that affect an organism or other specified system during its lifetime; the earth's life-support systems for us and for all other forms of life in effect another term for describing solar capital and earth capital.

environmental audit
an independent assessment of the current status of a party's compliance with applicable environmental requirements or of a party's environmental compliance policies, practices, and controls.

environmental degradation
a reduction of an ecosystem's or habitat's ability to support its natural biota. 2. Depletion or destruction of a potentially renewable resource such as soil, grassland, forest, or wildlife by using it at a faster rate than it is naturally replenished. If such use is continued, the resource can become non-renewable (on a human timescale) or non-existant. 3. Pollution, toxification, or other alteration of an environment that makes it less productive, hospit[illegible] able, or enjoyable.

■ environmental equity/ justice
equal protection from environmental hazards for individuals, groups, or communities regardless of race, ethnicity, or economic status. This applies to the development, implementation, and enforcement of environmental laws, regulations, and policies, and implies that no population of people should be forced to shoulder a disproportionate share of negative environmental impacts of pollution or environmental hazard due to a lack of political or economic strength levels.

■ environmental exposure
human exposure to pollutants originating from facility emissions. Threshold levels are not necessarily surpassed, but low-level chronic pollutant exposure is one of the most common forms of environmental exposure (See threshold level).

■ environmental fate
the destiny of a chemical or biological pollutant after release into the environment.

■ environmental fate data
data that Characterise a pesticide's fate in the ecosystem, considering factors that foster its degradation (light, water, microbes), pathways and resultant products.

■ environmental heterogeneity
the physical or temporal patchiness of the environment. Heterogeneity exists at all scales within natural communities, ranging from habitat differences between the top and underside of a leaf, to habitat patches created by treefalls within a forest, to the pattern of forests and grasslands within a region. The mosaic of habitat patches within an ecosystem is created by such disturbances as fire and storms; differences in microclimate, soils, and history; and both deterministic and random population variation. Patches in early stages of succession provide unique structural habitats and contain different species than those in late-successional stages do.

■ environmental indicator
a measurement, statistic or value that provides a proximate gauge or evidence of the effects of environmental management programs or of the state or condition of the environment.

■ environmental lien
a charge, security, or encumbrance on a property's title to secure payment of cost or debt arising from response actions, cleanup, or other remediation of hazardous substances or petroleum products.

■ environmental medium
a major environmental category that surrounds or contacts humans, animals, plants, and other organisms (e.g. surface water, ground water, soil or air) and through which chemicals or pollutants move. (See ambient medium, biological medium).

■ environmental site assessment
the process of determining whether contamination is present on a parcel of real property.

■ environmental sustainability
long-term maintenance of ecosystem components and functions for future generations.

■ environmental tobacco smoke
mixture of smoke from the burning end of a cigarette, pipe, or cigar and smoke exhaled by the smoker. (See passive smoking/secondhand smoke.).

■ environmental worldview
how individuals think the world works, what they think their role in the world should be, and what they believe is right and wrong environmental behaviour (i.e. ethics).

■ environmental/ecological risk
the potential for adverse effects on living organisms associated with pollution of the environment by effluents, emissions, wastes, or accidental chemical releases; energy use; or the depletion of natural resources.

■ **ephemeral**
an organism that has a short life cycle.

■ **epidemiology**
study of the patterns of disease or other harmful effects from toxic exposure within defined groups of people to find out why some people get sick and some do not.

■ **epilimnion**
upper waters of a thermally stratified lake subject to wind action.

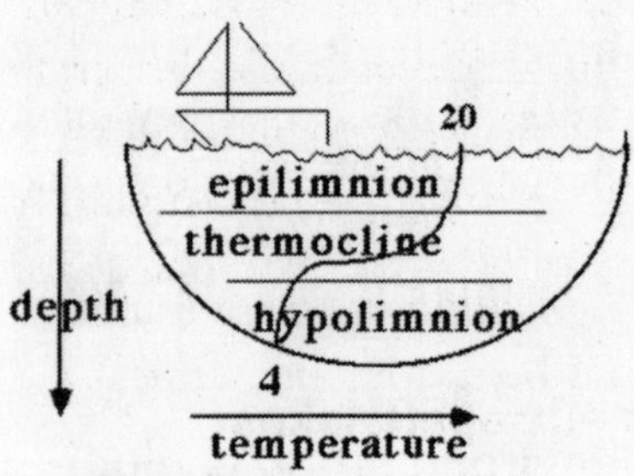

■ **epiphyte**
a plant that grows on another plant in a relation ship of commensalism.

■ **episode (pollution)**
an air pollution incident in a given area caused by a concentration of atmospheric pollutants under meteorological conditions that may result in a significant increase in illnesses or deaths. May also describe water pollution events or hazardous material spills.

■ **equator**
the imaginary boundary that divides the earth in half north and south.

■ **equilibrium**
in relation to radiation, the state at which the radioactivity of consecutive elements within a radioactive series is neither increasing nor decreasing.

■ **erid, acronym**
emerging and Re-emerging Infectious Diseases.

■ **erosion**
the wearing away of land surface by wind or water, intensified by land-clearing practices related to farming, residential or industrial development, road building, or logging.

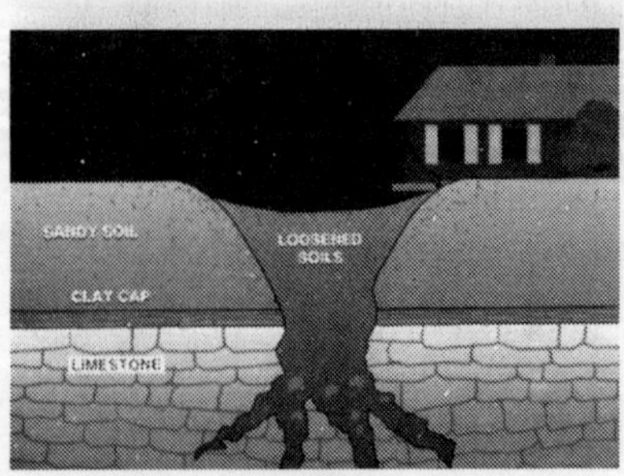

■ established treatment technologies

technologies for which cost and performance data are readily available. (See Innovative treatment technologies.).

■ estimated environmental concentration

the estimated pesticide concentration in an ecosystem.

■ estivate

to spend the summer in a sleeplike condition of partial or total inactivity.

■ estuary

region of interaction between rivers and near-shore ocean waters, where tidal action and river flow mix fresh and salt water. Such areas include bays, mouths of rivers, salt marshes, and lagoons. These brackish water ecosystems shelter and feed marine life, birds, and wildlife.

■ ethanol

an alternative automotive fuel derived from grain and corn; usually blended with gasoline to form gasohol.

■ ethnobotany

the study of indigenous knowledge bases regarding plants and their uses.

■ ethnopharmacology

the study of indigenous knowledge bases regarding medicines and how they are produced, as well as the medical practices, treatment protocols, etc. that utilize these medicines.

■ ethylene dibromide (EDB)

a chemical used as an agricultural fumigant and in certain industrial processes. Extremely toxic and found to be a carcinogen in laboratory animals,

■ eutrophic lakes

shallow, murky bodies of water with concentrations of plant nutrients causing excessive production of algae.

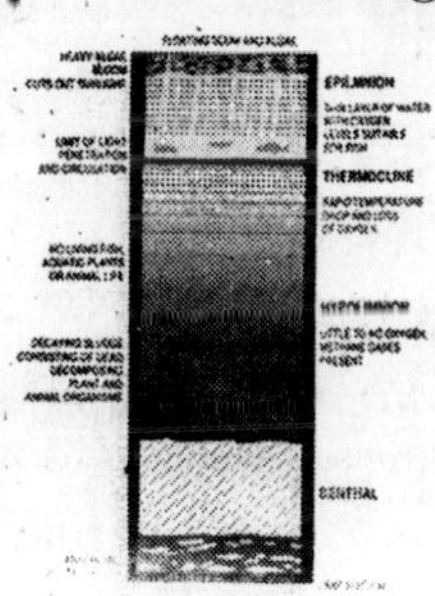

■ eutrophication

the slow aging process during which a lake, estuary, or bay evolves into a bog or marsh and eventually disappears. During the later stages of eutrophication the water body is choked by abundant plant life due to higher levels of nutritive compounds such as nitrogen and phosphorus. Human activities can accelerate the process.

■ evaporate

to change from a liquid to a gas as a result of being heated.

■ evapouration ponds

areas where sewage sludge is dumped and dried.

■ evapotranspiration

the loss of water from the soil both by evaporation and by transpiration from the plants growing in the soil.

■ evergreen

a plant whose needles or leaves remain green throughout the year.

■ evolution

any gradual change. Organic evolution is any genetic change in organisms from generation to generation.

■ ex situ conservation

a conservation method that entails the removal of germplasm resources (seed, pollen, sperm, individual organisms, from their original habitat or natural environment. Keeping components of biodiversity alive outside of their original habitat or natural environment.

■ exceedance

violation of the pollutant levels permitted by environmental protection standards.

■ exclusionary ordinance

zoning that excludes classes of persons or businesses from a particular neighbourhood or area.

■ exempted aquifer

underground bodies of water defined in the Underground Injection Control program as aquifers that are potential sources of drinking water though not being used as such, and thus exempted from regulations barring underground injection activities.

■ exotic species
a species that is not indigenous to a region.

■ experimental use permit
obtained by manufacturers for testing new pesticides or uses thereof whenever they conduct experimental field studies to support registration on 10 acres or more of land or one acre or more of water.

■ explosive limits
the amounts of vapour in the air that form explosive mixtures; limits are expressed as lower and upper limits and give the range of vapour concentrations in air that will explode if an ignition source is present.

■ exponential growth
growth in which some quantity, such as population size or economic output, increases by a fixed percentage of the whole in a given time; when the increase in quantity over a long enough time is plotted, this type of growth typically yields a curve shaped like the letter J.

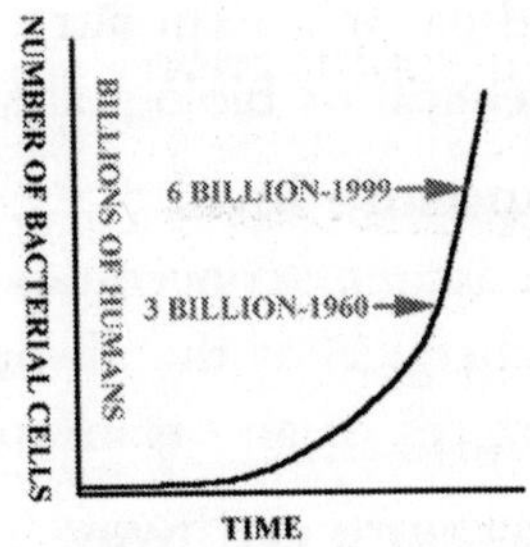

■ exposure
the amount of radiation or pollutant present in a given environment that represents a potential health threat to living organisms.

■ exposure assessment
identifying the pathways by which toxicants may reach individuals, estimating how much of a chemical an individual is likely to be exposed to, and estimating the number likely to be exposed.

■ exposure concentration
the concentration of a chemical or other pollutant representing a health threat in a given environment.

■ exposure indicator
a characteristic of the environment measured to provide evidence of the occurrence or magnitude of a response

indicator's exposure to a chemical or biological stress.

■ **exposure level**
the amount (concentration) of a chemical at the absorptive surfaces of an organism.

■ **exposure pathway**
the path from sources of pollutants via, soil, water, or food to man and other species or settings.

■ **exposure route**
the way a chemical or pollutant enters an organism after contact; i.e. by ingestion, inhalation, or dermal absorption.

■ **exposure-response relationship**
the relationship between exposure level and the incidence of adverse effects.

■ **extant**
a species that is still alive and reproducing. All species that currently live on earth are extant.

■ **extinct**
a species that is no longer living on earth. All representatives of the species are dead. All the species that once occupied the earth but are no longer living are extinct. We know of their existence through studying the fossil record . Compare to extant .

■ **extinction**
complete disappearance of a species from the earth. This happens when a species cannot adapt and successfully reproduce under new environmental conditions, when it evolves (through a process called radiation) into one or more new species, or when every member of the species is killed by overpredation, pollution, or other man-made causes.

■ **extraction procedure**
determining toxicity by a procedure which simulates leaching; if a certain concentration of a toxic substance can be leached from a waste, that waste is considered hazardous, i.e."EP Toxic."

■ **extraction well**
a discharge well used to remove groundwater or air.

■ **fabric filter**
a cloth device that catches dust particles from industrial emissions.

facultative bacteria

bacteria that can live under aerobic or anaerobic conditions.

fauna

all of the animals found in a given area.

feasibility study

1. analysis of the practicability of a proposal; e.g., a description and analysis of potential cleanup alternatives for a site such as one on the National Priorities List. The feasibility study usually recommends selection of a cost-effective alternative. It usually starts as soon as the remedial investigation is underway; together, they are commonly referred to as the "RI/FS". 2. A small-scale investigation of a problem to ascertain whether a proposed research approach is likely to provide useful data.

faecal coliform bacteria

bacteria found in the intestinal tracts of mammals. Their presence in water or sludge is an indicator of pollution and possible contamination by pathogens.

feedlot

a confined area for the controlled feeding of animals. Tends to concentrate large amounts of animal waste that cannot be absorbed by the soil and, hence, may be carried to nearby streams or lakes by rainfall runoff.

fen

a type of wetland that accumulates peat deposits. Fens are less acidic than bogs, deriving most of their water from groundwater rich in calcium and magnesium. (See wetlands.).

ferrous metals

magnetic metals derived from iron or steel; products made from ferrous metals include appliances, furniture, containers, and packaging like steel drums and barrels. Recycled products include processing tin/steel cans, strapping, and metals from appliances into new products.

fertilise

to join male sperm with a female egg.

fill

man-made deposits of natural soils or rock products and waste materials.

filling

depositing dirt, mud or other materials into aquatic areas to create more dry land, usually for agricultural or commercial development purposes, often with ruinous ecological consequences.

filter strip

strip or area of vegetation used for removing sediment, organic matter, and other pollutants from runoff and wastewater.

filtration

a treatment process, under the control of qualified operators, for removing solid (particulate) matter from water by means of porous media such as sand or a man-made filter; often used to remove particles that contain pathogens.

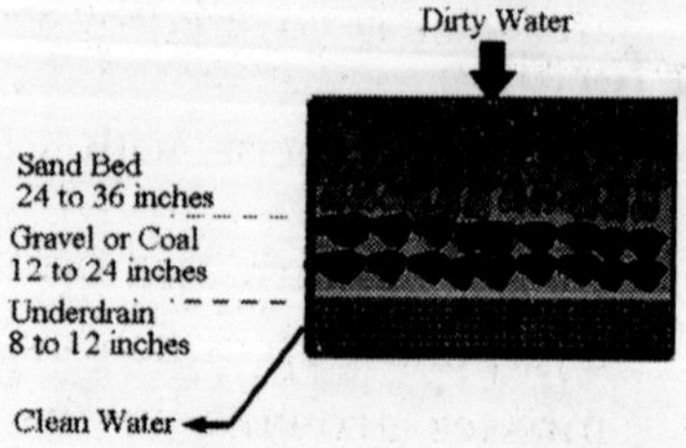

finding of no significant impact

a document prepared by a federal agency showing why a proposed action would not have a significant impact on the environment and thus would not require preparation of an Environmental Impact Statement. An FNSI is based on the results of an environmental assessment.

finished water

water is "finished" when it has passed through all the processes in a water treatment plant and is ready to be delivered to consumers.

first draw

the water that comes out when a tap is first opened, likely to have the highest level of lead contamination from plumbing materials.

first law of human ecology

we can never do merely one thing. Any intrusion into nature has numerous effects, many of which are unpredictable. For example, one classic dilemma is the case of behavioural biologists who observe their study subjects at close range Are the observed behaviours truly natural or are they influenced by the researcher's presence?

■ **first-order consumers**
animals that eat plants.

■ **fix a sample**
a sample is "fixed" in the field by adding chemicals that prevent water quality indicators of interest in the sample from changing before laboratory measurements are made.

■ **fixed-location monitoring**
sampling of an environmental or ambient medium for pollutant concentration at one location continuously or repeatedly.

■ **flammable**
any material that ignites easily and will burn rapidly.

■ **flare**
a control device that burns hazardous materials to prevent their release into the environment; may operate continuously or intermittently, usually on top of a stack.

■ **flash point**
the lowest temperature at which evaporation of a substance produces sufficient vapour to form an ignitable mixture with air.

■ **floc**
a clump of solids formed in sewage by biological or chemical action.

■ **flocculation**
process by which clumps of solids in water or sewage aggregate through biological or chemical action so they can be separated from water or sewage.

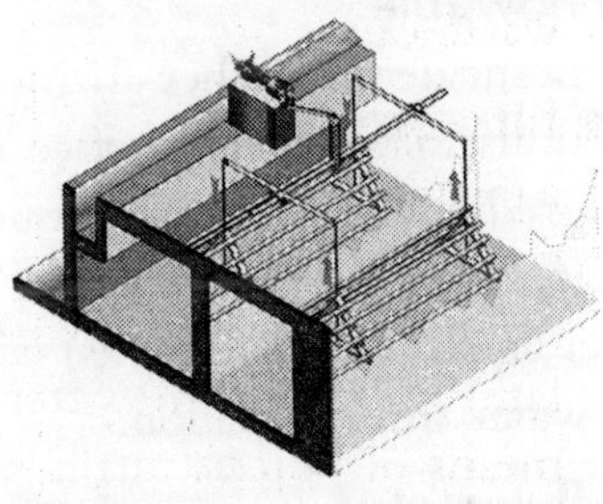

■ **floodplain**
the flat or nearly flat land along a river or stream or in a tidal area that is covered by water during a flood.

■ **floor layer**
a forest's sixth and bottom layer, made up of lichens and mosses growing in the re-

mains of fallen trees, branches, and leaves.

floor sweep
capture of heavier-than-air gases that collect at floor level.

flora
all the plants in a particular area.

flow rate
the rate, expressed in gallons -or liters-per-hour, at which a fluid escapes from a hole or fissure in a tank. Such measurements are also made of liquid waste, effluent, and surface water movement.

flowable
pesticide and other formulations in which the active ingredients are finely ground insoluble solids suspended in a liquid. They are mixed with water for application.

flowmeter
a gauge indicating the velocity of wastewater moving through a treatment plant or of any liquid moving through various industrial processes.

flue gas
the air coming out of a chimney after combustion in the burner it is venting. It can include nitrogen oxides, carbon oxides, water vapour, sulphur oxides, particles and many chemical pollutants.

flue gas desulphurisation
a technology that employs a sorbent, usually lime or limestone, to remove sulphur dioxide from the gases produced by burning fossil fuels. Flue gas desulphurisation is current state-of-the art technology for major SO_2 emitters, like power plants.

fluid
matter in a gas or liquid state.

fluidised
a mass of solid particles that is made to flow like a liquid by injection of water or gas is said to have been Fluidised. In water treatment, a bed of filter media is Fluidised by backwashing water through the filter.

fluidised bed incinerator
an incinerator that uses a bed of hot sand or other granular material to transfer heat directly to waste. Used mainly

for destroying municipal sludge.

flume

a natural or man-made channel that diverts water.

fluoridation

the addition of a chemical to increase the concentration of fluoride ions in drinking water to reduce the incidence of tooth decay.

fluorides

gaseous, solid, or dissolved compounds containing fluorine that result from industrial processes. Excessive amounts in food can lead to fluorosis.

fluorocarbons (FC's)

any of a number of organic compounds analogous to hydrocarbons in which one or more hydrogen atoms are replaced by fluorine. Once used in the United States as a propellant for domestic aerosols, they are now found mainly in coolants and some industrial processes. FCs containing chlorine are called chlorofluorocarbons (CFCs). They are believed to be modifying the ozone layer in the stratosphere, thereby allowing more harmful solar radiation to reach the Earth's surface.

flush

1. to open a cold-water tap to clear out all the water which may have been sitting for a long time in the pipes. In new homes, to flush a system means to send large volumes of water gushing through the unused pipes to remove loose particles of solder and flux. 2. To force large amounts of water through a system to clean out piping or tubing, and storage or process tanks.

flux

1. a flowing or flow. 2. A substance used to help metals fuse together. fly ash noncombustible residual particles expelled by flue gas. fogging applying a pesticide by rapidly heating the liquid chemical so that it forms very fine droplets that resemble smoke or fog. Used to destroy mosquitoes, black flies, and similar pests.

food chain

figure of speech describing the dependence of heterotrophs on other organisms for food, progressing in a series beginning with primary producers (plants) and ending with the largest carnivores. The food chain is used as a figurative image for educational purposes only real trophic systems resemble webs rather than chains. See food web.

Marine Food Chain

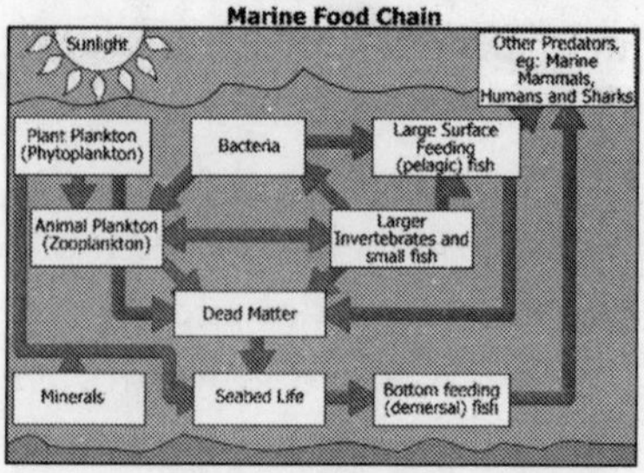

■ food processing waste

food residues produced during agricultural and industrial operations.

■ food waste

uneaten food and food preparation wastes from residences and commercial establishments such as grocery stores, restaurants, and produce stands, institutional cafeterias and kitchens, and industrial sources like employee lunchrooms.

■ food web

all of the interlinked food chains in a community or an ecosystem.

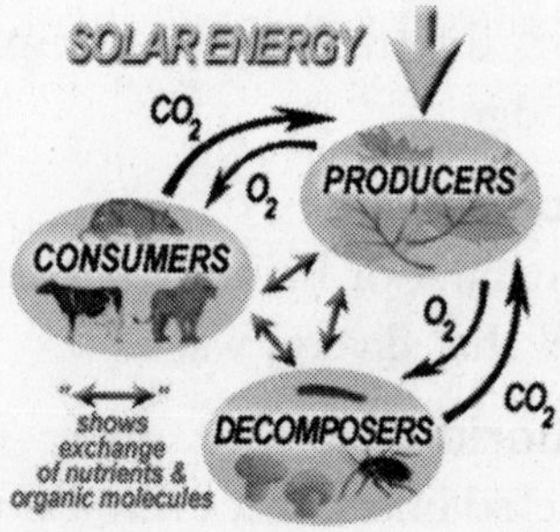

■ forest

a biome whose main vegetation consists of large groups of trees that usually grow close enough together that their tops touch, shading the ground.

■ formaldehyde

a colourless, pungent, and irritating gas, CH_2O, used chiefly as a disinfectant and preservative and in synthesising other compounds like resins.

■ formulation

the substances comprising all active and inert ingredients in a pesticide.

■ fossil

a remnant, impression, mineralised mould, amber encasement, or other trace of a once-living organism. Technically, anything that once lived and has been permanently preserved is a fossil, but the most common usage implies great age. This common usage of fossil generally refers to the mineralised remains or impressions, preserved in stone (almost always sedimentary rock), of extinct organisms from past geologic ages.

■ **fossil fuel**

products of partial of complete decomposition of plants and animals that occur as crude oil, coal, natural gas, or heavy oils as a result of exposure to heat and pressure in earth's crust over millions of years.

■ **fossil record**

the cumulative taxonomic information and historical perspective provided by the wealth and diversity of fossils and related geologic data stored in the earth's crust.

■ **fracture**

a break in a rock formation due to structural stresses; e.g. faults, shears, joints, and planes of fracture cleavage.

■ **free product**

a petroleum hydrocarbon in the liquid free or non-aqueous phase.

■ **freeboard**

1.vertical distance from the normal water surface to the top of a confining wall. 2. Vertical distance from the sand surface to the underside of a trough in a sand filter.

■ **fresh water**

water that generally contains less than 1,000 milligrams-per-litre of dissolved solids.

■ **friable**

capable of being crumbled, pulverised, or reduced to powder by hand pressure.

■ **friable asbestos**

any material containing more than one-percent asbestos, and that can be crumbled or reduced to powder by hand pressure. (May include previously non-friable material which becomes broken or damaged by mechanical force.).

■ frugivore

an animal that eats fruit.

■ fuel efficiency

the proportion of energy released by fuel combustion that is converted into useful energy.

■ fuel switching

1. a precombustion process whereby a low-sulphur coal is used in place of a higher sulphur coal in a power plant to reduce sulphur dioxide emissions. 2. Illegally using leaded gasoline in a motor vehicle designed to use only unleaded.

■ fugitive emissions

emissions not caught by a capture system.

■ fume

tiny particles trapped in vapour in a gas stream.

■ fumigant

a pesticide vaporised to kill pests. Used in buildings and greenhouses.

■ fungicide

pesticides which are used to control, deter, or destroy fungi.

■ fungistat

a chemical that keeps fungi from growing.

■ fungus (fungi)

moulds, mildews, yeasts, mushrooms, and puffballs, a group of organisms lacking in chlorophyll (i.e. are not photosynthetic) and which are usually non-mobile, filamentous, and multicellular. Some grow in soil, others attach themselves to decaying trees and other plants whence they obtain nutrients. Some are pathogens, others Stabilise sewage and digest composted waste.

■ furrow irrigation

irrigation method in which water travels through the field by means of small channels between each groups of rows.

■ future liability

refers to potentially responsible parties' obligations to

pay for additional response activities beyond those specified in the Record of Decision or Consent Decree.

■ **game fish**
species like trout, salmon, or bass, caught for sport. Many of them show more sensitivity to environmental change than "rough" fish.

■ **garbage**
animal and vegetable waste resulting from the handling, storage, sale, preparation, cooking, and serving of foods.

■ **gas chromatograph/ mass spectrometer**
instrument that identifies the molecular composition and concentrations of various chemicals in water and soil samples.

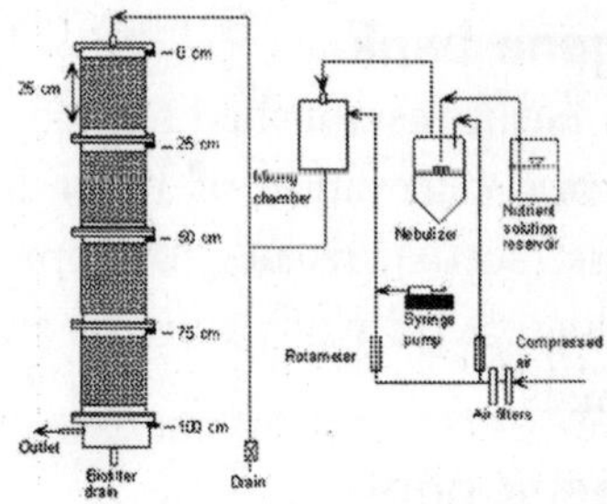

■ **gasahol**
mixture of gasoline and ethanol derived from fermented agricultural products containing at least nine percent ethanol. Gasohol emissions contain less carbon monoxide than those from gasoline.

■ **gasification**
conversion of solid material such as coal into a gas for use as a fuel.

■ **gasoline volatility**
the property of gasoline whereby it evaporates into a vapour. Gasoline vapour is a mixture of volatile organic compounds.

■ **gene**
the functional unit of heredity; the part of the DNA molecule that encodes a single enzyme or structural protein unit.

■ gene bank
a facility established for the *ex situ* conservation of individuals (seeds), tissues, or reproductive cells of plants or animals.

■ gene pool
the sum total of all the genes that exist among all the individuals of a species.

■ general permit
a permit applicable to a class or category of dischargers.

■ generalist
organism which can survive under a wide variety of conditions, and does not specialize to live under any particular set of circumstances.

■ generator
1. a facility or mobile source that emits pollutants into the air or releases hazardous waste into water or soil. 2. any person, by site, whose act or process produces regulated medical waste or whose act first causes such waste to become subject to regulation. Where more than one person (e.g. doctors with separate medical practices) are located in the same building, each business entity is a separate generator.

■ genetic diversity
variety among individuals within a species or, more specifically, the variety in the DNA of a species. See also "alleles."

■ genetic engineering
the artificial transfer of specific genes from one organism to another.

■ genotype
the set of genes possessed by an individual organism.

■ geographic information system (GIS)
a computer system designed for storing, manipulating, analysing, and displaying data in a geographic context.

■ geologic time scale
occurring at such a slow pace, or at such infrequent intervals, as to be imperceptible to humans. 2. Occurring in a prehuman era. 3. The whole of earth's history, as opposed to the very recent period when humans have walked the earth. One common and effective means of conceptualising the disparity between the geologic

time scale and the human time scale is the "calendar year history model," wherein the entire history of the planet is condensed into a single calendar year. In this model, human ancestors do not appear until late December and Homo sapiens does not arise until the last second before midnight on December 31st.

■ geological log

a detailed description of all underground features (depth, thickness, type of formation) discovered during the drilling of a well.

■ geology

the branch of science that deals with the earth's history, particularly its physical history, as recorded in the substrate and the fossil record .

■ geophysical log

a record of the structure and composition of the earth encountered when drilling a well or similar type of test hold or boring.

■ geopolitics

the study of the influence of such factors as geographyatural resources, economics, and demography on the politics (especially the foreign policy) of nations.

■ geothermal/ground source heat pump

these heat pumps are underground coils to transfer heat from the ground to the inside of a building.

■ germicide

any compound that kills disease-causing microorganisms.

■ germplasm

the genetic material, especially its specific molecular and chemical constitution, that comprises the physical basis of the inherited qualities of an organism.

■ giardia lamblia

protozoan in the faeces of humans and animals that can cause severe gastrointestinal ailments. It is a common contaminant of surface waters.

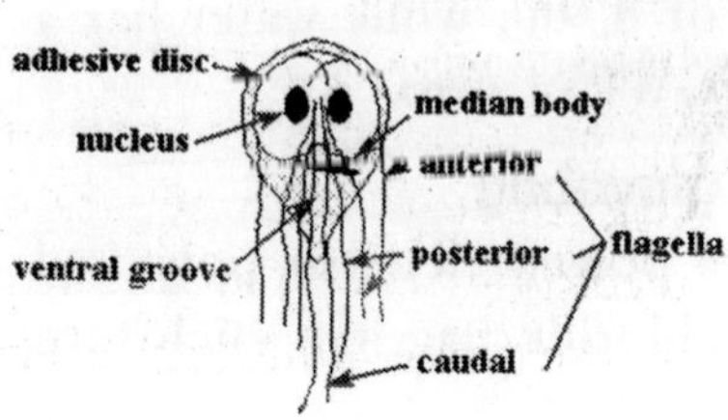

■ glass containers
for recycling purposes, containers like bottles and jars for drinks, food, cosmetics and other products. When being recycled, container glass is generally separated into colour categories for conversion into new containers, construction materials or fiberglass insulation.

■ global warming
the term given to the possibility that Earth's atmosphere is gradually warming because of the greenhouse effect of carbon dioxide and other gases. Global warming is thought by many to be the most serious global environmental issue facing our society.

■ global warming potential
the ratio of the warming caused by a substance to the warming caused by a similar mass of carbon dioxide. CFC-12, for example, has a GWP of 8,500, while water has a GWP of zero.

■ glovebag
a polyethylene or polyvinyl chloride bag-like enclosure affixed around an asbestos-containing source (most often thermal system insulation) permitting the material to be removed while Minimising release of airborne fibers to the surrounding atmosphere.

■ gooseneck
a portion of a water service connection between the distribution system water main and a meter. Sometimes called a pigtail.

■ grab sample
a single sample collected at a particular time and place that represents the composition of the water, air, or soil only at that time and place.

■ grain loading
the rate at which particles are emitted from a pollution source. Measurement is made by the number of grains per cubic foot of gas emitted.

■ granular activated carbon treatment
a filtering system often used in small water systems and individual homes to remove organics. Also used by municipal water treatment plants. GAC can be highly effective

in lowering elevated levels of radon in water.

■ **grasscycling**
source reduction activities in which grass clippings are left on the lawn after mowing.

■ **grassed waterway**
natural or constructed watercourse or outlet that is shaped or graded and established in suitable vegetation for the disposal of runoff water without erosion.

■ **grassland**
region in which the climate is dry for long periods of the summer, and freezes in the winter. Grasslands are Characterised by grasses and other erect herbs, usually without trees or shrubs. Grasslands occur in the dry temperate interiors of continents, and first appeared in the Miocene.

■ **grassroots**
people or society at a local level, rather than at the center of major political activity.

■ **gray water**
domestic wastewater composed of wash water from kitchen, bathroom, and laundry sinks, tubs, and washers.

■ **green revolution**
refers to the development and introduction of new varieties of wheat and rice (mainly) that increased yields per acre dramatically in some countries.

■ **greenhouse**
a structure, usually made of glass or clear plastic, that provides a protected, controlled environment for raising plants indoors.

■ **greenhouse effect**
the warming of the Earth's atmosphere attributed to a build-

up of carbon dioxide or other gases; some scientists think that this build-up allows the sun's rays to heat the Earth, while making the infra-red radiation atmosphere opaque to infra-red radiation, thereby preventing a counterbalancing loss of heat.

■ **greenhouse gas**

a gas, such as carbon dioxide or methane, which contributes to potential climate change.

■ **grinder pump**

a mechanical device that shreds solids and raises sewage to a higher elevation through pressure sewers.

■ **gross alpha/beta particle activity**

the total radioactivity due to alpha or beta particle emissions as inferred from measurements on a dry sample.

■ **gross power-generation potential**

the installed power generation capacity that landfill gas can support.

■ **gross primary productivity**

the rate at which an ecosystem's producers capture and store a given amount of chemical energy as biomass in a given length of time.

■ **ground cover**

plants grown to keep soil from eroding.

■ **ground water**

the supply of fresh water found beneath the Earth's surface, usually in aquifers, which supply wells and springs. Because ground water is a major source of drinking water, there is growing concern over contamination from leaching agricultural or industrial pollutants or leaking underground storage tanks.

■ **ground water under the direct influence of surface water**

any water beneath the surface of the ground with

1. significant occurence of insects or other microorganisms, algae, or large-diameter pathogens.

2. significant and relatively rapid shifts in water characteristics such as turbidity, temperature, conductivity, or pH which closely correlate to climatological or surface water

conditions. Direct influence is determined for individual sources in accordance with criteria established by a statc.

■ ground-penetrating radar

a geophysical method that uses high frequency electromagnetic waves to obtain subsurface information.

■ ground-water discharge

ground water entering near coastal waters which has been contaminated by landfill leachate, deep well injection of hazardous wastes, septic tanks, etc.

■ guest

organism living on or in a host; a parasite.

■ guild

a group of organisms that share a common food resource.

■ gully erosion

severe erosion in which trenches are cut to a depth greater than 30 centimeters (a foot). Generally, ditches deep enough to cross with farm equipment are considered gullies.

■ habitat

the environment in which an organism lives. Habitat can also refer to the organisms and physical environment in a particular place.

■ habitat indicator

a physical attribute of the environment measured to Characterise conditions necessary to support an organism, population, or community in the absence of pollutants; e.g. salinity of estuarine waters or substrate type in streams or lakes.

■ half-life

1. the time required for a pollutant to lose one-half of its original concentration or example, the biochemical half-life of DDT in the environment is 15 years.

2. the time required for half of the atoms of a radioactive element to undergo self-transmutation or decay (half-life of radium is 1620 years).

3. the time required for the elimination of half a total dose from the body.

■ **halogen**

a type of incandescent lamp with higher energy-efficiency that standard ones.

■ **halon**

bromine-containing compounds with long atmospheric lifetimes whose breakdown in the stratosphere causes depletion of ozone. Halons are used in firefighting.

■ **halophile**

organism which lives in areas of high salt concentration. These organisms must have special adaptations to permit them to survive under these conditions.

■ **hard water**

alkaline water containing dissolved salts that interfere with some industrial processes and prevent soap from sudsing.

■ **hauler**

garbage collection company that offers complete refuse removal service; many will also collect recyclables.

■ **hazard**

1. potential for radiation, a chemical or other pollutant to cause human illness or injury. 2. In the pesticide program, the inherent toxicity of a compound. Hazard identification of a given substances is an informed judgment based on verifiable toxicity data from animal models or human studies.

■ **hazard assessment**

evaluating the effects of a stressor or determining a margin of safety for an organism by comparing the concentration, which causes toxic effects with an estimate of exposure to the organism.

■ **hazard evaluation**

a component of risk evaluation that involves gathering and evaluating data on the types of health injuries or diseases that may be produced by a chemical and on the conditions of exposure under which such health effects are produced.

■ hazard identification
determining if a chemical or a microbe can cause adverse health effects in humans and what those effects might be.

■ hazard quotient
the ratio of estimated site-specific exposure to a single chemical from a site over a specified period to the estimated daily exposure level, at which no adverse health effects are likely to occur.

■ hazard ratio
a term used to compare an animal's daily dietary intake of a pesticide to its LD 50 value. A ratio greater than 1.0 indicates that the animal is likely to consume an a dose amount which would kill 50 percent of animals of the same species.

■ hazardous air pollutants
air pollutants which are not covered by ambient air quality standards but which, as defined in the Clean Air Act, may present a threat of adverse human health effects or adverse environmental effects. Such pollutants include asbestos, beryllium, mercury, benzene, coke oven emissions, radionuclides, and vinyl chloride.

■ hazardous waste
any solid, liquid, or containerised gas that can catch fire easily, is corrosive to skin tissue or metals, is unstable and can explode or release toxic fumes, or has harmful concentrations of one or more toxic materials that can leach out.

■ hazardous waste landfill
an excavated or engineered site where hazardous waste is deposited and covered.

■ hazardous waste minimisation
reducing the amount of toxicity or waste produced by a facility via source reduction or environmentally sound recycling.

■ hazards analysis
procedures used to (1) identify potential sources of release of hazardous materials from fixed facilities or transportation accidents; (2) determine the vulnerability of a

geographical area to a release of hazardous materials; and (3) compare hazards to determine which present greater or lesser risks to a community.

■ hazards identification

providing information on which facilities have extremely hazardous substances, what those chemicals are, how much there is at each facility, how the chemicals are stored, and whether they are used at high temperatures.

■ headspace

the vapour mixture trapped above a solid or liquid in a sealed vessel.

■ health advisory level

a non-regulatory health-based reference level of chemical traces (usually in ppm) in drinking water at which there are no adverse health risks when ingested over various periods of time. Such levels are established for one day, 10 days, long-term and life.time exposure periods. They contain a wide margin of safety.

■ heat island effect

a "dome" of elevated temperatures over an urban area caused by structural and pavement heat fluxes, and pollutant emissions.

■ heat pump

an electric device with both heating and cooling capabilities. It extracts heat from one medium at a lower (the heat source) temperature and transfers it to another at a higher temperature (the heat sink), thereby cooling the first and warming the second. (See geothermal, water source heat pump).

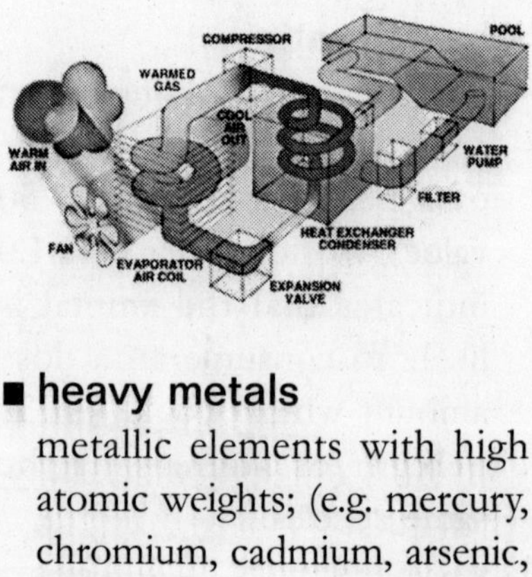

■ heavy metals

metallic elements with high atomic weights; (e.g. mercury, chromium, cadmium, arsenic, and lead); can damage living things at low concentrations and tend to accumulate in the food chain.

■ heptachlor

an insecticide that was banned on some food products in 1975 and in all of them 1978. It was allowed for use in seed treatment until 1983. More recently it was found in milk and other dairy products in Arkansas and Missouri where dairy cattle were illegally fed treated seed.

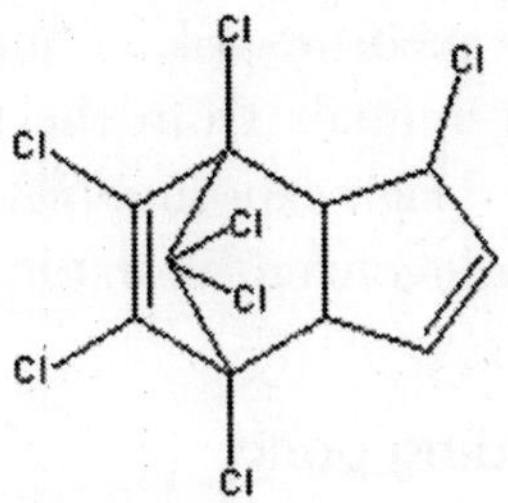

■ herb layer

a forest's fifth layer, found close to the ground and containing plants such as flowers, grasses, ferns, seedling trees, and shrubs.

■ herbicide

a chemical pesticide designed to control or destroy plants, weeds, or grasses.

■ herbivore

literally, an organism that eats plants or other autotrophic organisms. The term is used primarily to describe animals.

■ heterotroph

literally, "eats others." An organism that must consume other organisms to fuel its metabolism. Animals, including humans, are heterotrophs. Heterotrophic.

■ hibernate

to spend the winter in a sleeplike condition of partial or total inactivity.

■ high end exposure (dose) estimate

an estimate of exposure, or dose level received anyone in a defined population that is greater than the 90th percentile of all individuals in that population, but less than the exposure at the highest percentile in that population. A high end risk descriptor is an estimate of the risk level for such individuals. Note that risk is based on a combination of exposure and susceptibility to the stressor.

■ high intensity discharge

a generic term for mercury vapour, metal halide, and high pressure sodium lamps and fixtures.

■ high-density polyethylene

a material used to make plastic bottles and other products

that produces toxic fumes when burned.

■ **highest dose tested**
the highest dose of a chemical or substance tested in a study.

■ **high-level nuclear waste facility**
plant designed to handle disposal of used nuclear fuel, high-level radioactive waste, and plutonium waste.

■ **high-level radioactive waste (HLRW)**
waste generated in core fuel of a nuclear reactor, found at nuclear reactors or by nuclear fuel reprocessing; is a serious threat to anyone who comes near the waste without shielding.

■ **high-line jumpers**
pipes or hoses connected to fire hydrants and laid on top of the ground to provide emergency water service for an isolated portion of a distribution system.

■ **high-risk community**
a community located within the vicinity of numerous sites of facilities or other potential sources of environmental exposure/health hazards which may result in high levels of exposure to contaminants or pollutants.

■ **high-to-low-dose extrapolation**
the process of prediction of low exposure risk to humans and animals from the measured high-exposure-high-risk data involving laboratory animals.

■ **holding pond**
a pond or reservoir, usually made of earth, built to store polluted runoff.

■ **holding time**
the maximum amount of time a sample may be stored before analysis.

■ **hollow stem auger drilling**
conventional drilling method that uses augurs to penetrate the soil. As the augers are rotated, soil cuttings are conveyed to the ground surface via augur spirals. DP tools can be used inside the hollow augers.

■ **homeowner water system**
any water system which supplies piped water to a single residence.

■ **homogeneous area**
in accordance with Asbestos Hazard and Emergency Response Act (AHERA) definitions, an area of surfacing materials, thermal surface insulation, or miscellaneous material that is uniform in colour and texture.

■ **hood capture efficiency**
ratio of the emissions captured by a hood and directed into a control or disposal device, expressed as a percent of all emissions.

■ **host**
an organism on or in which a parasite lives and whose support of the parasite often leads to its own injury.

■ **hot desert**
a desert with hot daytime temperatures for most of the year.

■ **household hazardous waste**
hazardous products used and disposed of by residential as opposed to industrial consumers. Includes paints, stains, varnishes, solvents, pesticides, and other materials or products containing volatile chemicals that can catch fire, react or explode, or that are corrosive or toxic.

■ **household waste (domestic waste)**
solid waste composed of garbage and rubbish, which normally originates in a private home or apartment house. Domestic waste may contain a significant amount of toxic or hazardous waste.

■ **human capital**
physical and mental talents of people used to produce, distribute, and sell an economic good.

■ **human equivalent dose**
a dose which, when administered to humans, produces an effect equal to that produced by a dose in animals.

■ **human exposure evaluation**
describing the nature and size of the population exposed to a substance and the magni-

tude and duration of their exposure.

■ **human health risk**
the likelihood that a given exposure or series of exposures may have damaged or will damage the health of individuals.

■ **human time scale**
occurring within a short enough time frame that the event can be perceived, remembered, and recounted by humans through oral traditions, written histories, or other mechanisms of human memory.

■ **hybrid**
the offspring of two parents from separate (though closely related) species. Usually sterile, though occassionally able to breed back into one of the parent lines. A hybrid can almost never produce viable offspring when mated with another hybrid. A common example is a mule, which is produced by breeding a horse with a donkey (note that the horse must be the mother, due to the large size of the foal). Hybridisation is fairly common among wind-pollinated plants, while Hybridisation is quite uncommon among higher animals.

■ **hybridisation**
crossing of individuals from genetically different strains, populations, or species.

■ **hydraulic conductivity**
the rate at which water can move through a permeable medium. (i.e. the coefficient of permeability.).

■ **hydraulic gradient**
in general, the direction of groundwater flow due to changes in the depth of the water table.

■ **hydrocarbons**
chemical compounds that consist entirely of carbon and hydrogen.

■ **hydrogen sulphide (H_2S)**
gas emitted during organic decomposition. Also a by-product of oil refining and burning. Smells like rotten eggs and, in heavy concentration, can kill or cause illness.

■ **hydrogeological cycle**
the natural process recycling water from the atmosphere

down to (and through) the earth and back to the atmosphere again.

■ hydrogeology
the geology of ground water, with particular emphasis on the chemistry and movement of water.

■ hydrologic cycle
biogeochemical cycle that collects, purifies, and distributes the earth's fixed supply of water, from the environment to living organisms and then back to the environment.

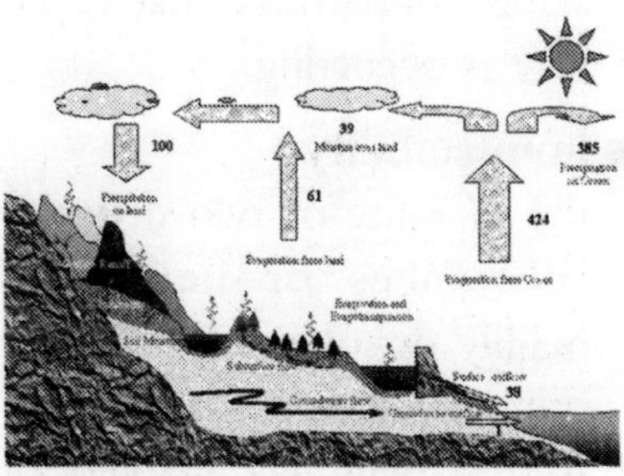

■ hydrology
the science dealing with the properties, distribution, and circulation of water.

■ hydrolysis
the decomposition of organic compounds by interaction with water.

■ hydronic
a ventilation system using heated or cooled water pumped through a building.

■ hydrophilic
having a strong affinity for water.

■ hydrophobic
having a strong aversion for water.

■ hydropneumatic
a water system, usually small, in which a water pump is automatically controlled by the pressure in a compressed air tank.

■ hydrosphere
the earth's liquid water (oceans, lakes other bodies of surface water, and underground water), the earth's frozen water (polar ice caps, floating ice caps, and ice in soil known as permafrost), and small amounts of water vapour in the atmosphere.

■ hypersensitivity diseases
diseases characterised by allergic responses to pollutants; diseases most clearly associated with indoor air quality are

asthma, rhinitis, and pneumonic hypersensitivity.

■ **hypolimnion**
bottom waters of a thermally stratified lake. The hypolimnion of a eutrophic lake is usually low or lacking in oxygen.

■ **hypoxia/hypoxic waters**
waters with dissolved oxygen concentrations of less than 2 ppm, the level generally accepted as the minimum required for most marine life to survive and reproduce.

■ **ignitable**
capable of burning or causing a fire.

■ **IM240**
a high-tech, transient dynamometer automobile emissions test that takes up to 240 seconds.

■ **IMHOFF CONE**
a clear, cone-shaped container used to measure the volume of settleable solids in a specific volume of water.

■ **immediately dangerous to life and health**
the maximum level to which a healthy individual can be exposed to a chemical for 30 minutes and escape without suffering irreversible health effects or impairing symptoms. Used as a "level of concern." (See level of concern.).

■ **imminent hazard**
one that would likely result in unreasonable adverse effects on humans or the environment or risk unreasonable hazard to an endangered species during the time required for a pesticide registration cancellation proceeding.

■ **imminent threat**
a high probability that exposure is occurring.

■ **immiscibility**
the inability of two or more substances or liquids to readily dissolve into one another, such as soil and water. Immiscibility The inability of two or more substances or liquids to readily dissolve into one another, such as soil and water.

■ **impermeable**
not easily penetrated. The property of a material or soil that does not allow, or allows only with great difficulty, the

movement or passage of water.

■ imports

municipal solid waste and recyclables that have been transported to a state or locality for processing or final disposition (but that did not originate in that state or locality).

■ impoundment

a body of water or sludge confined by a dam, dike, flood gate, or other barrier.

■ in situ

in its original place; unmoved unexcavated; remaining at the site or in the subsurface.

■ in situ conservation

a conservation method that attempts to preserve the genetic integrity of gene resources by conserving them within the evolutionary dynamic ecosystems of the original habitat or natural environment.

■ in vitro

storage of plant or animal germplasm in tissue-culture form in glass containers.

■ in vivo

testing or action inside an organism.

■ inbreeding

a mating system involving the mating or breeding of closely related individuals, the most extreme form of which is self-fertilisation. It is used to "fix" economically useful genetic traits in genetically improved populations; however, it also can result in fixation of deleterious recessive alleles.

■ inbreeding depression

a reduction in fitness or vigour as a result of fixation of deleterious, recessive alleles from consistent inbreeding in a normally out breeding population.

■ incident command post

a facility located at a safe distance from an emergency site, where the incident commander, key staff, and technical representatives can make decisions and deploy emergency manpower and equipment.

■ incident command system (ICS)

the organisational arrangement wherein one person, normally the Fire Chief of the impacted district, is in charge of an integrated, comprehensive emergency response organisation and the emergency incident site, backed by an Emergency Operations Centre staff with resources, information, and advice.

■ incineration

a treatment technology involving destruction of waste by controlled burning at high temperatures; e.g., burning sludge to remove the water and reduce the remaining residues to a safe, non-burnable ash that can be disposed of safely on land, in some waters, or in underground locations.

■ incineration at sea

disposal of waste by burning at sea on specially-designed incinerator ships.

■ incinerator

a furnace for burning waste under controlled conditions.

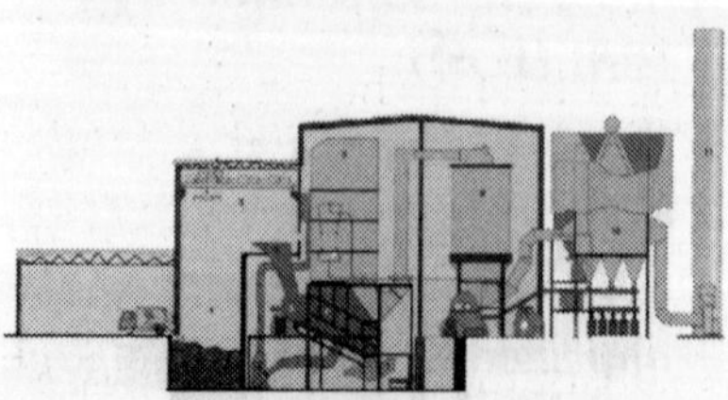

■ incompatible waste

a waste unsuitable for mixing with another waste or material because it may react to form a hazard.

■ indicator

in biology, any biological entity or processes, or community whose characteristics show the presence of specific environmental conditions. 2. In chemistry, a substance that shows a visible change, usually of colour, at a desired point in a chemical reaction. 3.A device that indicates the result of a measurement; e.g. a pressure gauge or a moveable scale.

■ indicator species

species that serve as early warnings that a community or an ecosystem is being degraded. Fish and amphibians make particularly excellent indicator species. Large predators (those generally at the apex of the food pyramid) are also good indicators in many habitats.

■ indigenous peoples

people whose ancestors inhabited a place or country

when persons from another culture or ethnic background arrived on the scene and dominated them through conquest, settlement, or other means and who today live more in conformity with their own social, economic, and cultural customs and traditions than with those of the country of which they now form a part.

■ indirect discharge
introduction of pollutants from a non-domestic source into a publicly owned waste-treatment system. Indirect dischargers can be commercial or industrial facilities whose wastes enter local sewers.

■ indirect source
any facility or building, property, road or parking area that attracts motor vehicle traffic and, indirectly, causes pollution.

■ indoor air
the breathable air inside a habitable structure or conveyance.

■ indoor air pollution
chemical, physical, or biological contaminants in indoor air.

■ indoor climate
temperature, humidity, lighting, air flow and noise levels in a habitable structure or conveyance. Indoor climate can affect indoor air pollution.

■ industrial pollution prevention
combination of industrial source reduction and toxic chemical use substitution.

■ industrial process waste
residues produced during manufacturing operations.

■ industrial sludge
semi-liquid residue or slurry remaining from treatment of industrial water and wastewater.

■ industrial source reduction
practices that reduce the amount of any hazardous substance, pollutant, or contaminant entering any waste stream or otherwise released into the environment. Also reduces the threat to public health and the environment associated with such releases. Term includes equipment or technology modifications, substitution of raw materials,

and improvements in housekeeping, maintenance, training or inventory control.

■ **industrial waste**

unwanted materials from an industrial operation; may be liquid, sludge, solid, or hazardous waste.

■ **inert ingredient**

pesticide components such as solvents, carriers, dispersants, and surfactants that are not active against target pests. Not all inert ingredients are innocuous.

■ **inertial separator**

a device that uses centrifugal force to separate waste particles.

■ **infectious agent**

any organism, such as a pathogenic virus, parasite, or bacterium, that is capable of invading body tissues, multiplying, and causing disease.

■ **infectious waste**

hazardous waste capable of causing infections in humans, including contaminated animal waste; human blood and blood products; isolation waste, pathological waste; and discarded sharps (needles, scalpels or broken medical instruments).

■ **infiltration**

1.the penetration of water through the ground surface into sub-surface soil or the penetration of water from the soil into sewer or other pipes through defective joints, connections, or manhole walls. 2. The technique of applying large volumes of waste water to land to penetrate the surface and percolate through the underlying soil. (See percolation.).

■ **infiltration gallery**

a sub-surface groundwater collection system, typically shallow in depth, constructed with open-jointed or perforated pipes that discharge collected water into a watertight chamber from which the water is pumped to treatment facilities and into the distribution system. Usually located close to streams or ponds.

■ **infiltration rate**

the quantity of water that can enter the soil in a specified time interval.

■ inflow
entry of extraneous rain water into a sewer system from sources other than infiltration, such as basement drains, manholes, storm drains, and street washing.

■ influent
water, wastewater, or other liquid flowing into a reservoir, basin, or treatment plant.

■ information file
in the Superfund program, a file that contains accurate, up-to-date documents on a Superfund site. The file is usually located in a public building (school, library, or city hall) convenient for local residents.

■ ingestion
the intake of water or food particles by "swallowing" them, taking them into the body cavity or into a vacuole. Contrast with absorption.

■ inhalable particles
all dust capable of entering the human respiratory tract.

■ initial compliance period (water)
the first full three-year compliance period which begins at least 18 months after promulgation.

■ injection well
a well into which fluids are injected for purposes such as waste disposal, improving the recovery of crude oil, or solution mining.

■ injection zone
a geological formation receiving fluids through a well.

■ in-line filtration
pre-treattment method in which chemicals are mixed by the flowing water; commonly used in pressure filtration installations. Eliminates need for flocculation and sedimentation.

■ innovative technologies
new or inventive methods to treat effectively hazardous waste and reduce risks to human health and the environment.

■ innovative treatment technologies
technologies whose routine use is inhibited by lack of data on performance and cost.

inoculum

1.bacteria or fungi injected into compost to start biological action. 2. a medium containing organisms, usually bacteria or a virus, that is introduced into cultures or living organisms.

inorganic chemicals

chemical substances of mineral origin, not of basically carbon structure.

insecticide

a pesticide compound specifically used to kill or prevent the growth of insects.

in-situ flushing

introduction of large volumes of water, at times supplemented with cleaning compounds, into soil, waste, or ground water to flush hazardous contaminants from a site.

in-situ oxidation

technology that oxidises contaminants dissolved in ground water, converting them into insoluble compounds.

in-situ stripping

treatment system that removes or "strips" volatile organic compounds from contaminated ground or surface water by forcing an airstream through the water and causing the compounds to evapourate.

in-situ vitrification

technology that treats contaminated soil in place at extremely high temperatures, at or more than 3000 degrees Fahrenheit.

insolation

the amount of solar energy that reaches the earth.

inspection and maintenance

1. activities to ensure that vehicles' emission controls work properly. 2. Also applies to wastewater treatment plants and other anti-pollution facilities and processes.

institutional waste

waste generated at institutions such as schools, libraries, hospitals, prisons, etc.

instream use

water use taking place within a stream channel; e.g., hydroelectric power generation, navigation, water quality improvement, fish propagation, recreation.

insulator
a material that does not easily gain or lose energy.

integrated exposure assessment
cumulative summation (over time) of the magnitude of exposure to a toxic chemical in all media.

integrated pest management
a mixture of chemical and other, non-pesticide, methods to control pests.

integrated waste management
using a variety of practices to handle municipal solid waste; can include source reduction, recycling, incineration, and landfilling.

intellectual property right
a right enabling an inventor to exclude imitators from the market for a limited time.

interceptor sewers
large sewer lines that, in a combined system, control the flow of sewage to the treatment plant. In a storm, they allow some of the sewage to flow directly into a receiving stream, thus keeping it from overflowing onto the streets. Also used in separate systems to collect the flows from main and trunk sewers and carry them to treatment points.

interface
the common boundary between two substances such as a water and a solid, water and a gas, or two liquids such as water and oil.

interfacial tension
the strength of the film separating two immiscible fluids (e.g. oil and water) measured in dynes per, or millidynes per centimeter.

internal dose
in exposure assessment, the amount of a substance penetrating the absorption barriers (e.g. skin, lung tissue, gastrointestinal tract) of an organism through either physical or biological processes. (See absorbed dose).

interspecific competition
members of two or more species trying to use the same lim-

ited resources in an ecosystem.

■ interstate carrier water supply

a source of water for drinking and sanitary use on planes, buses, trains, and ships operating in more than one state. These sources are federally regulated.

■ interstate waters

waters that flow across or form part of state or international boundaries; e.g. the Great Lakes, the Mississippi River, or coastal waters.

■ interstitial monitoring

the continuous surveillance of the space between the walls of an underground storage tank.

■ intraspecific competition

two or individual organisms of a single species trying to use the same limited resources in an ecosystem.

■ intrastate product

pesticide products once registered by states for sale and use only in the state. All intrastate products have been converted to full federal registration or cancelled.

■ introduced species

a species occurring in an area outside of its historically known natural range as a result of intentional or accidental dispersal by human activities. Also known as alien species.

■ inversion

a layer of warm air that prevents the rise of cooling air and traps pollutants beneath it; can cause an air pollution episode.

■ ion

an electrically charged atom or group of atoms.

■ ion exchange treatment

a common water-softening method often found on a large scale at water purification plants that remove some organics and radium by adding calcium oxide or calcium hydroxide to increase the pH to a level where the metals will precipitate out.

■ ionisation chamber

a device that measures the intensity of Ionising radiation.

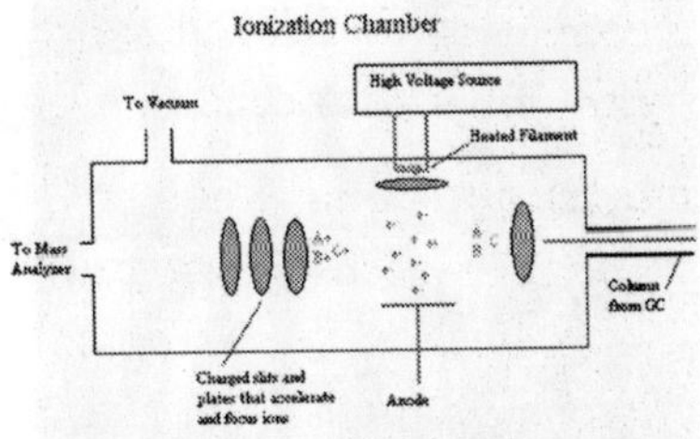

■ **ionising radiation**
radiation that can strip electrons from atoms; e.g. alpha, beta, and gamma radiation.

■ **irradiated food**
food subject to brief radioactivity, usually gamma rays, to kill insects, bacteria, and Mould, and to permit storage without refrigeration.

■ **irradiation**
exposure to radiation of wavelengths shorter than those of visible light (gamma, x-ray, or ultra- violet), for medical purposes, to Sterilise milk or other foodstuffs, or to induce polymerisation of monomers or vulcanisation of rubber.

■ **irreversible effect**
effect Characterised by the inability of the body to par tially or fully repair injury caused by a toxic agent.

■ **irrigation**
applying water or wastewater to land areas to supply the water and nutrient needs of plants.

■ **irrigation efficiency**
the amount of water stored in the crop root zone compared to the amount of irrigation water applied.

■ **irrigation return flow**
surface and subsurface water which leaves the field following application of irrigation water.

■ **irritant**
a substance that can cause irritation of the skin, eyes, or respiratory system. Effects may be acute from a single high level exposure, or chronic from repeated low-level exposures to such compounds as chlorine, nitrogen dioxide, and nitric acid.

■ **isoconcentration**
more than one sample point exhibiting the same isolate concentration.

■ **isopleth**
the line or area represented by an isoconcentration.

■ **isotope**
a variation of an element that has the same atomic number of protons but a different weight because of the number of neutrons. Various isotopes of the same element may have different radioactive behaviours, some are highly unstable..

■ **isotropy**
the condition in which the hydraulic or other properties of an aquifer are the same in all directions.

■ **jar test**
a laboratory procedure that simulates a water treatment plant's coagulation/flocculation units with differing chemical doses, mix speeds, and settling times to estimate the minimum or ideal coagulant dose required to achieve certain water quality goals.

■ **karst**
a geologic formation of irregular limestone deposits with sinks, underground streams, and caverns.

■ **kelp**
underwater forests of tall, brown algae that grow in cool coastal waters.

■ **kelp forest**
marine ecosystem dominated by large kelps. These forests are restricted to cold and temperate waters, and are most common along the western coasts of continents. Kelp forests first appeared in the Miocene.

■ **keystone species**
a species whose loss from an ecosystem would cause a greater than average change in other species populations or ecosystem processes.

■ **kinetic energy**
energy possessed by a moving object or water body.

■ **kinetic rate coefficient**
a number that describes the rate at which a water constituent such as a biochemical oxygen demand or dissolved oxygen rises or falls, or at which an air pollutant reacts.

kwashiorkor

type of malnutrition that occurs in infants and very young children when they are weaned from mother's milk to a starchy diet low in protein.

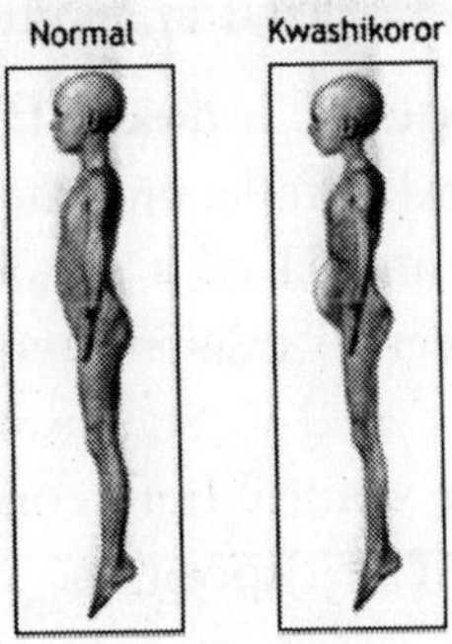

laboratory animal studies

investigations using animals as surrogates for humans.

lagoon

1. a shallow pond where sunlight, bacterial action, and oxygen work to purify wastewater; also used for storage of wastewater or spent nuclear fuel rods.
2. shallow body of water, often separated from the sea by coral reefs or sandbars.

land application

discharge of wastewater onto the ground for treatment or reuse.

land disposal restrictions

rules that require hazardous wastes to be treated before disposal on land to destroy or immobilise hazardous constituents that might migrate into soil and ground water.

land farming (of waste)

a disposal process in which hazardous waste deposited on or in the soil is degraded naturally by microbes.

landfills

1. sanitary landfills are disposal sites for non-hazardous solid wastes spread in layers, compacted to the smallest practical volume, and covered by material applied at the end of each operating day. 2. secure chemical landfills are disposal sites for hazardous waste, selected and designed to minimise the chance of release of hazardous substances into the environment.

landraces

a crop cultivar or animal breed that evolved with and has been genetically improved by traditional agriculturalists, but has not been influenced by modern breeding practices.

■ landscape

the traits, patterns, and structure of a specific geographic area, including its biological composition, its physical environment, and its anthropogenic or social patterns. An area where interacting ecosystems are grouped and repeated in similar form.

■ landscape characterisation

documentation of the traits and patterns of the essential elements of the landscape.

■ landscape ecology

the study of the distribution patterns of communities and ecosystems, the ecological processes that affect those patterns, and changes in pattern and process over time.

■ landscape indicator

a measurement of the landscape, calculated from mapped or remotely sensed data, used to describe spatial patterns of land use and land cover across a geographic area. Landscape indicators may be useful as measures of certain kinds of environmental degradation such as forest fragmentation.

■ land-use planning

process for deciding the best present and future use of each parcel of land in an area.

■ Langelier Index (LI)

an index reflecting the equilibrium pH of a water with respect to calcium and alkalinity; used in Stabilising water to control both corrosion and scale deposition.

Figure 3: An LSI Of 2.5 Coincides With A Calcite Saturation Level of 150 in Lower TDS Waters

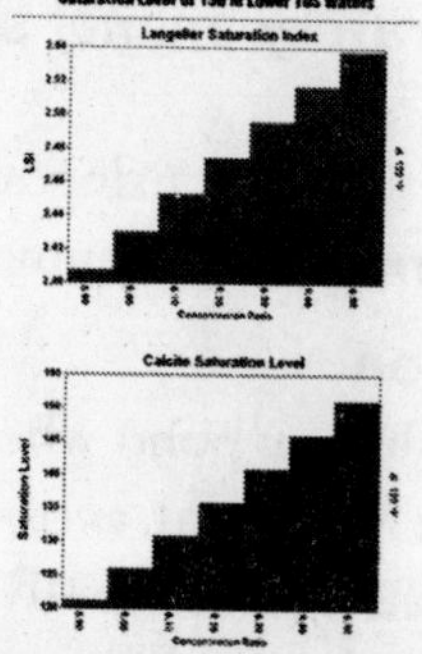

■ large quantity generator

person or facility generating more than 2200 pounds of hazardous waste per month. Such generators produce about 90 percent of the nation's hazardous waste, and are subject to all RCRA requirements.

■ **large water system**
a water system that services more than 50,000 customers.

■ **laser induced fluorescence**
a method for measuring the relative amount of soil and/or groundwater with an in-situ sensor.

■ **late successional plant species**
mostly trees that can tolerate shade and that form a relatively stable complex forest community.

■ **latency**
time from the first exposure of a chemical until the appearance of a toxic effect.

■ **lateral sewers**
pipes that run under city streets and receive the sewage from homes and businesses, as opposed to domestic feeders and main trunk lines.

■ **laterite**
product of rock decomposition with high iron and aluminium hydroxide content. Generally bright red to deep orange in colour. 2. Land, usually in the tropics, baked by the sun after deforestation removes the protective and restorative forest layer above the soil. Abiotic hardpack ground, red in colour. Normal soil microbiotic community, as well as macrobiotic flora and fauna, are absent. Prone to extensive erosion due to lack of plant cover. Lateritised hillsides have contributed to several devastating and deadly landslides in tropical countries.

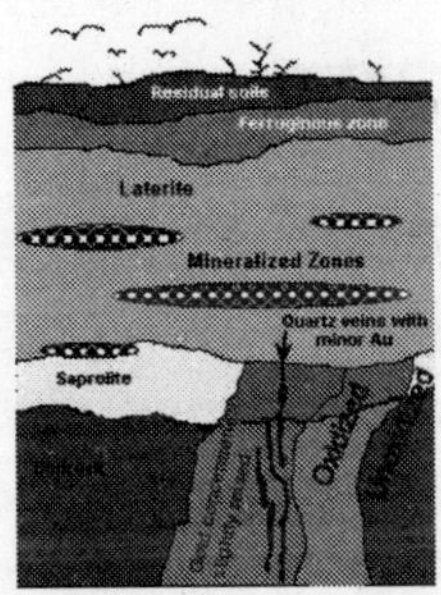

■ **lateritisation**
the process of turning formerly healthy soils into laterite. What becomes of tropical forest lands when deforested and left exposed to the elements. The ground becomes extremely hard and cannot be penetrated by germinating forest seeds, so recolonisation is slow or absent.

laundering weir
sedimention basin overflow weir.

LD 50/ lethal dose
the dose of a toxicant or microbe that will kill 50 percent of the test organisms within a designated period. The lower the LD 50, the more toxic the compound.

Ldlo
lethal dose low; the lowest dose in an animal study at which lethality occurs.

leachate
a mixture of rainwater and other liquids that comes from garbage.

leachate collection system
a system that gathers leachate and pumps it to the surface for treatment.

leaching
the process by which soluble constituents are dissolved and filtered through the soil by a percolating fluid. (See leachate.).

lead (PB)
a heavy metal that is hazardous to health if breathed or swallowed. Its use in gasoline, paints, and plumbing compounds has been sharply restricted or eliminated by federal laws and regulations. (See heavy metals.).

lead service line
a service line made of lead which connects the water to the building inlet and any lead fitting connected to it.

legionella
a genus of bacteria, some species of which have caused a type of pneumonia called Legionaires Disease.

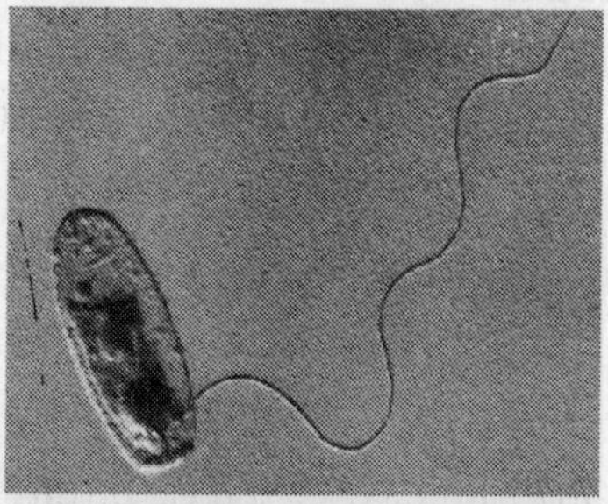

level of concern (LOC)
the concentration in air of an extremely hazardous substance above which there may be serious immediate health effects to anyone exposed to it for short periods.

■ **lichen**
a symbiotic relationship between a fungus and a moss. The moss does most of the work, producing sugars for the lichen's collective metabolic pathways. Lichen are generally low-growing, vary in colour from bright orange or yellow to grey or black, and are often found growing on rocks and tree bark. An easy mnemonic to assist recall of the nature of a lichen's symbiosis is "A fungus took a liking to a moss, and now they live together."

■ **life cycle of a product**
all stages of a product's development, from extraction of fuel for power to production, marketing, use, and disposal.

■ **life form**
characteristic structure of a plant or animal.

■ **lifetime average daily dose**
figure for estimating excess lifetime cancer risk.

■ **lifetime exposure**
total amount of exposure to a substance that a human would receive in a lifetime (usually assumed to be 70 years).

■ **lift**
in a sanitary landfill, a compacted layer of solid waste and the top layer of cover material.

■ **lifting station**
(See pumping station.).

■ **Light Non-Aqueous Phase Liquid (LNAPL)**
a non-aqueous phase liquid with a specific gravity less than 1.0. Because the specific gravity of water is 1.0, most LNAPLs float on top of the water table. Most common petroleum hydrocarbon fuels and lubricating oils are LNAPLs.

■ **light-emitting diode**
a long-lasting illumination technology used for exit signs which requires very little power.

■ **limestone scrubbing**

use of a limestone and water solution to remove gaseous stack-pipe sulphur before it reaches the atmosphere.

■ **limit of detection (LoD)**

the minimum concentration of a substance being analysed test that has a 99 percent probability of being identified.

■ **limited degradation**

an environmental policy permitting some degradation of natural systems but terminating at a level well beneath an established health standard.

■ **limiting factor**

a condition whose absence or excessive concentration, is incompatible with the needs or tolerance of a species or population and which may have a negative influence on their ability to thrive.

■ **limnology**

the study of river system ecology and life.

■ **lindane**

a pesticide that causes adverse health effects in domestic water supplies and is toxic to freshwater fish and aquatic life.

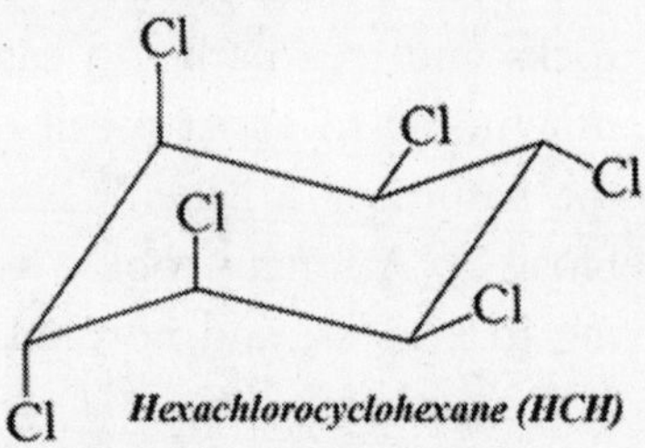

Hexachlorocyclohexane (HCH)

■ **liner**

1. a relatively impermeable barrier designed to keep leachate inside a landfill. Liner materials include plastic and dense clay. 2. An insert or sleeve for sewer pipes to prevent leakage or infiltration.

■ **lipid solubility**

the maximum concentration of a chemical that will dissolve in fatty substances. Lipid soluble substances are insoluble in water. They will very selectively disperse through the environment via uptake in living tissue.

■ liquefaction
changing a solid into a liquid.

■ liquid
flowing freely like water.

■ liquid injection incinerator
commonly used system that relies on high pressure to prepare liquid wastes for incineration by breaking them up into tiny droplets to allow easier combustion.

■ lithology
mineralogy, grain size, texture, and other physical properties of granular soil, sediment, or rock.

■ litter
leaf litter, or forest litter, is the detritus of fallen leaves and bark which accumulate in forests.

■ littoral zone
1. that portion of a body of fresh water extending from the shoreline lakeward to the limit of occupancy of rooted plants. 2. A strip of land along the shoreline between the high and low water levels.

■ low density polyethylene
plastic material used for both rigid containers and plastic film applications.

■ low emissivity windows
new window technology that lowers the amount of energy loss through windows by inhibiting the transmission of radiant heat while still allowing sufficient light to pass through.

■ low NOx burners
one of several combustion technologies used to reduce emissions of Nitrogen Oxides (NO_x.).

■ lower detection limit
the smallest signal above background noise an instrument can reliably detect.

■ lower explosive limit
the concentration of a compound in air below which the mixture will not catch on fire.

■ lowest acceptable daily dose
the largest quantity of a chemical that will not cause a toxic effect, as determined by animal studies.

■ lowest observed ad-

verse effect level
the lowest level of a stressor that causes statistically and biologically significant differences in test samples as compared to other samples subjected to no stressor.

■ **macropores**
secondary soil features such as root holes or desiccation cracks that can create significant conduits for movement of NAPL and dissolved contaminants, or vapour-phase contaminants.

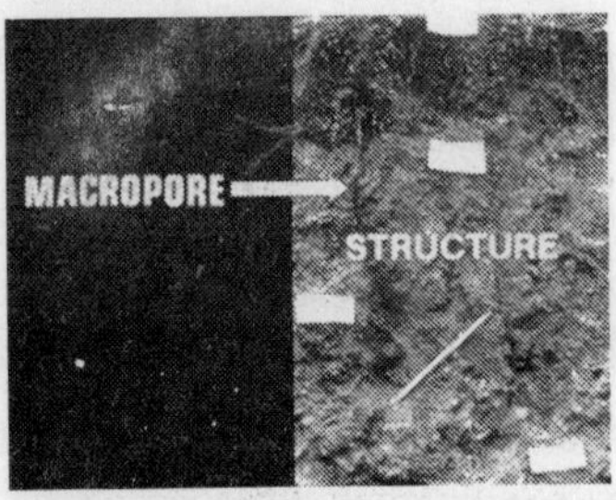

■ **magma**
molten, or melted, rock within the earth.

■ **magnetic separation**
use of magnets to separate ferrous materials from mixed municipal waste stream.

■ **major stationary sources**
term used to determine the applicability of Prevention of Significant Deterioration and new source regulations. In a non-attainment area, any stationary pollutant source with potential to emit more than 100 tons per year is considered a major stationary source. In PSD areas the cutoff level may be either 100 or 250 tons, depending upon the source.

■ **majors**
larger publicly owned treatment works (POTWs) with flows equal to at least one million gallons per day (mgd) or servicing a population equivalent to 10,000 persons; certain other POTWs having significant water quality impacts.

■ **managerial controls**
methods of nonpoint source pollution control based on decisions about managing agricultural wastes or application times or rates for agrochemicals.

■ **mandatory recycling**
programs which by law require consumers to separate trash so that some or all recyclable materials are recovered

for recycling rather than going to landfills.

■ **manifest system**
tracking of hazardous waste from "cradle-to-grave" (generation through disposal) with accompanying documents known as manifests.(See cradle to grave.).

■ **man-made (anthropogenic) beta particle and photon emitters**
all radionuclides emitting beta particles and/or photons listed in Maximum Permissible Body Burdens and Maximum Permissible Concentrations of Radonuclides in Air and Water for Occupational Exposure.

■ **mantle**
the part of the earth between the crust and the core.

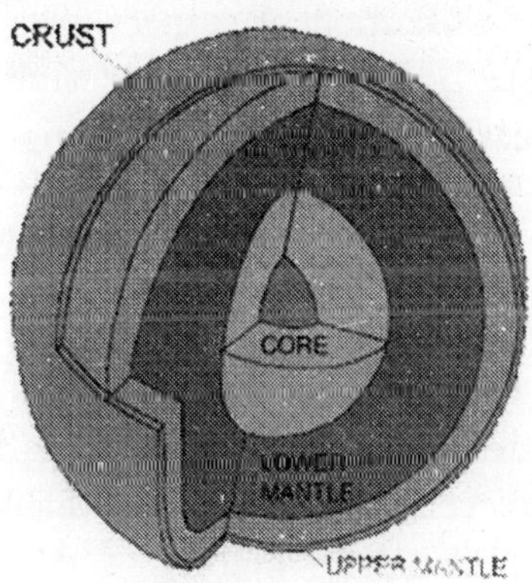

■ **manual separation**
hand sorting of recyclable or compostable materials in waste.

■ **manufacturer's formulation**
a list of substances or component parts as described by the maker of a coating, pesticide, or other product containing chemicals or other substances.

■ **manufacturing use product**
any product intended (labeled) for formulation or repackaging into other pesticide products.

■ **marasmus**
nutritional-deficiency disease caused by a diet that does not have enough calories and protein to maintain good health.

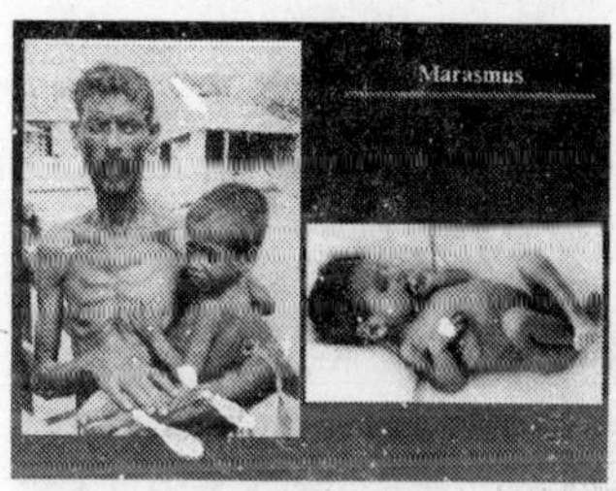

■ **margin of exposure**
the ratio of the no-observed adverse-effect-level to the estimated exposure dose.

■ **margin of safety**

maximum amount of exposure producing no measurable effect in animals (or studied humans) divided by the actual amount of human exposure in a population.

■ **marine**

refers to the ocean.

■ **marine life**

plants and animals of the ocean.

■ **marine sanitation device**

any equipment or process installed on board a vessel to receive, retain, treat, or discharge sewage.

■ **marsh**

a type of wetland that does not accumulate appreciable peat deposits and is dominated by herbaceous vegetation. Marshes may be either fresh or saltwater, tidal or nontidal. (See wetlands.).

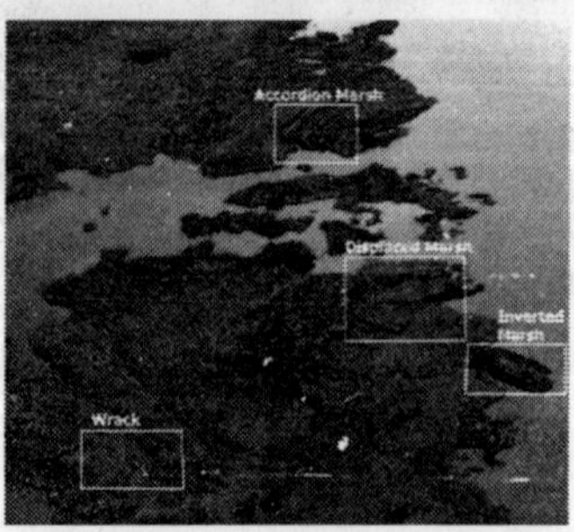

■ **mass extinction**

a catastrophic, widespread—often global—event in which major groups of species are wiped out over a relatively short period when compared to normal (background) extinction rates. There have been five major mass extinctions, of natural causes (in at least one case due to an asteroid impacting the earth), in the earth's history. We are now entering a sixth great mass extinction, this time of unnatural causes. human activities.

■ **material category**

in the asbestos program, broad classification of materials into thermal surfacing insulation, surfacing material, and miscellaneous material.

■ **material type**

classification of suspect material by its specific use or application; e.g., pipe insulation, fireproofing, and floor tile.

■ **materials recovery facility**

a facility that processes residentially collected mixed recyclables into new products available for market.

maximally (or most) exposed individual
the person with the highest exposure in a given population.

maximum acceptable toxic concentration
for a given ecological effects test, the range (or geometric mean) between the No Observable Adverse Effect Level and the Lowest Observable Adverse Effects Level.

maximum contaminant level
the maximum permissible level of a contaminant in water delivered to any user of a public system. MCLs are enforceable standards.

maximum tolerated dose
the maximum dose that an animal species can tolerate for a major portion of its lifetime without significant impairment or toxic effect other than carcinogenicity.

measure of effect/ measurement endpoint
a measurable characteristic of ecological entity that can be related to an assessment endpoint; e.g. a laboratory test for eight species meeting certain requirements may serve as a measure of effect for an assessment endpoint, such as survival of fish, aquatic, invertebrate or algal species under acute exposure.

measure of exposure
a measurable characteristic of a stressor (such as the specific amount of mercury in a body of water) used to help quantify the exposure of an ecological entity or individual organism.

mechanical aeration
use of mechanical energy to inject air into water to cause a waste stream to absorb oxygen.

mechanical separation
using mechanical means to separate waste into various components.

mechanical turbulence
random irregularities of fluid motion in air caused by build ings or other nonthermal, processes.

media
specific environments—air, water, soil—which are the

subject of regulatory concern and activities.

■ **median lethal dose (LD50)**
amount of a toxic material per unit body weight of test animals that kills half the test population in a certain time.

■ **medical surveillance**
a periodic comprehensive review of a worker's health status; acceptable elements of such surveillance program are listed in the Occupational Safety and Health Administration standards for asbestos.

■ **medical waste**
any solid waste generated in the diagnosis, treatment, or immunisation of human beings or animals, in research pertaining thereto, or in the production or testing of biologicals, excluding hazardous waste identified .

■ **medium-size water system**
a water system that serves 3,300 to 50,000 customers.

■ **meniscus**
the curved top of a column of liquid in a small tube.

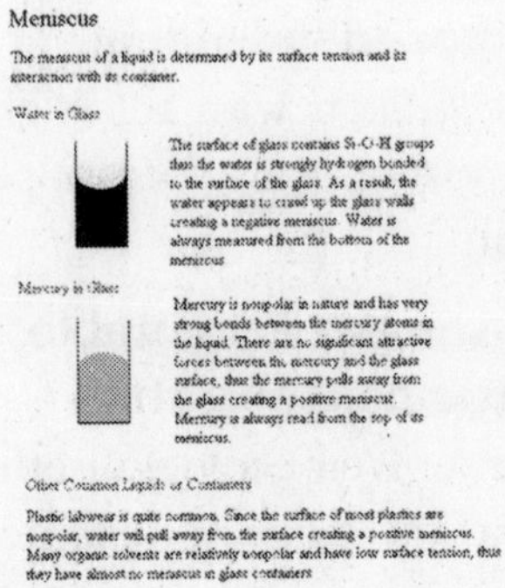

■ **mercury (Hg)**
heavy metal that can accumulate in the environment and is highly toxic if breathed or swallowed. (See heavy metals.).

■ **mesotrophic**
reservoirs and lakes which contain moderate quantities of nutrients and are moderately productive in terms of aquatic animal and plant life.

■ **metabolites**
any substances produced by biological processes, such as those from pesticides.

■ **metalimnion**
the middle layer of a thermally stratified lake or reservoir. In this layer there is a rapid decrease in temperature with depth. Also called thermocline.

■ meteorologist
a scientist who studies the weather.

■ methane
a colourless, nonpoisonous, flammable gas created by anaerobic decomposition of organic compounds. A major component of natural gas used in the home.

■ methanol
an alcohol that can be used as an alternative fuel or as a gasoline additive. It is less volatile than gasoline; when blended with gasoline it lowers the carbon monoxide emissions but increases hydrocarbon emissions. Used as pure fuel, its emissions are less ozone-forming than those from gasoline. Poisonous to humans and animals if ingested.

■ methyl orange alkalinity
a measure of the total alkalinity in a water sample in which the colour of methyl orange reflects the change in level.

■ microbial growth
the amplification or multiplication of microorganisms such as bacteria, algae, diatoms, plankton, and fungi.

■ microbial pesticide
a microorganism that is used to kill a pest, but is of minimum toxicity to humans.

■ microclimate
1. localised climate conditions within an urban area or neighbourhood. 2. The climate around a tree or shrub or a stand of trees.

■ microenvironmental method
a method for sequentially assessing exposure for a series of microenvironments that can be approximated by constant concentrations of a stressor.

■ microenvironments
well-defined surroundings such as the home, office, or kitchen that can be treated as uniform in terms of stressor concentration.

■ microorganisms
living organisms so small they can only be seen through a microscope.

■ midnight zone
the area of the ocean beneath the twilight zone, extending from 3,000 feet (1,000 m) down to the ocean floor, where only about 1 percent of marine life can survive.

■ migrate
to move from one place to another.

■ million-gallons per day (MGD)
a measure of water flow.

■ minimisation
a comprehensive program to minimise or eliminate wastes, usually applied to wastes at their point of origin.

■ minimum viable population
the smallest isolated population having a good chance of surviving for a given number of years despite the foreseeable effects of demographic, environmental, and genetic events and natural catastrophes. (The probability of persistence and the time of persistence are often taken to be 99 percent and 1000 years, respectively.).

■ mining of an aquifer
withdrawal over a period of time of ground water that exceeds the rate of recharge of the aquifer.

■ mining waste
residues resulting from the extraction of raw materials from the earth.

■ minors
publicly owned treatment works with flows less than 1 million gallons per day. (See majors.).

■ miscellaneous ACM
interior asbestos-containing building material or structural components, members or fixtures, such as floor and ceiling tiles; does not include surfacing materials or thermal system insulation.

■ miscellaneous materials
interior building materials on structural components, such as floor or ceiling tiles.

■ miscible liquids
two or more liquids that can be mixed and will remain mixed under normal conditions.

■ missed detection
the situation that occurs when a test indicates that a tank is "tight" when in fact it is leaking.

■ mist
liquid particles measuring 40 to 500 micrometers (pm), are formed by condensation of vapour. By comparison, fog particles are smaller than 40 micrometers (pm).

■ mitigation
measures taken to reduce adverse impacts on the environment.

■ mixed glass
recovered container glass not sorted into categories (e.g. colour, grade).

■ mixed liquor
a mixture of activated sludge and water containing organic matter undergoing activated sludge treatment in an aeration tank.

■ mixed metals
recovered metals not sorted into categories such as aluminium, tin, or steel cans or ferrous or non-ferrous metals.

■ mixed municipal waste
solid waste that has not been sorted into specific categories (such as plastic, glass, yard trimmings, etc.).

■ mixed paper
recovered paper not sorted into categories such as old magazines, old newspapers, old corrugated boxes, etc.

■ mixed plastic
recovered plastic unsorted by category.

■ mobile incinerator systems
hazardous waste incinerators that can be transported from one site to another.

■ mobile source
any non-stationary source of air pollution such as cars, trucks, motorcycles, buses, airplanes, and locomotives.

■ model plant
a hypothetical plant design used for developing economic, environmental, and energy impact analyses as support for regulations or regulatory guidelines; first step in exploring the economic impact of a potential NSPS.

■ **modified bin method**

way of calculating the required heating or cooling for a building based on determining how much energy the system would use if outdoor temperatures were within a certain temperature interval and then multiplying the energy use by the time the temperature interval typically occurs.

■ **modified source**

the enlargement of a major stationary pollutant sources is often referred to as modification, implying that more emissions will occur.

■ **moisture content**

1.the amount of water lost from soil upon drying to a constant weight, expressed as the weight per unit of dry soil or as the volume of water per unit bulk volume of the soil. For a fully saturated medium, moisture content indicates the porosity.

2. Water equivalent of snow on the ground; an indicator of snowmelt flood potential.

■ **molecule**

the smallest part of a substance that has all the characteristics of the substance.

■ **molten salt reactor**

a thermal treatment unit that rapidly heats waste in a heat-conducting fluid bath of carbonate salt.

■ **monitoring**

periodic or continuous surveillance or testing to determine the level of compliance with statutory requirements and/or pollutant levels in various media or in humans, plants, and animals.

■ **monitoring well**

1.a well used to obtain water quality samples or measure groundwater levels. 2. A well drilled at a hazardous waste management facility or Superfund site to collect ground-water samples for the purpose of physical, chemical, or biological analysis to determine the amounts, types, and distribution of contaminants in the groundwater beneath the site.

monoclonal antibodies

1. man-made (anthropogenic) clones of a molecule, produced in quantity for medical or research purposes.
2. molecules of living organisms that selectively find and attach to other molecules to which their structure conforms exactly. This could also apply to equivalent activity by chemical molecules.

monocropping

the act of planting and maintaining a monoculture .

monoculture

cultivation of a single crop, usually on a large area of land. This unnatural agricultural system generally requires the use of large quantities of artificial fertilise rs, herbicides, pesticidesematocides, and other pest control efforts. Even with these chemical aids, monocultures are prone to disease outbreaks and pest infestations, largely due to the genetic homogeneity of such systems.

monomictic

lakes and reservoirs, which are relatively deep, do not freeze over during winter, and undergo a single stratification and mixing cycle during the year (usually in the fall).

montreal protocol

treaty, signed in 1987, governs stratospheric ozone protection and research, and the production and use of ozone-depleting substances. It provides for the end of production of ozone-depleting substances such as CFCS. Under the Protocol, various research groups continue to assess the ozone layer. The Multilateral Fund provides resources to developing nations to promote the transition to ozone-safe technologies.

morbidity

rate of disease incidence.

mortality

death rate.

most probable number

an estimate of microbial density per unit volume of water sample, based on probability theory.

mountain

a biome of high ground with various types of vegetation depending on the elevation.

muck soils
earth made from decaying plant materials.

mudballs
round material that forms in filters and gradually increases in size when not removed by backwashing.

mulch
a layer of material (wood chips, straw, leaves, etc.) placed around plants to hold moisture, prevent weed growth, and enrich or Sterilise the soil.

multi-media approach
joint approach to several environmental media, such as air, water, and land.

multiple chemical sensitivity
a diagnostic label for people who suffer multi-system illnesses as a result of contact with, or proximity to, a variety of airborne agents and other substances.

multiple use
use of land for more than one purpose; e.g., grasing of livestock, watershed and wildlife protection, recreation, and timber production. Also applies to use of bodies of water for recreational purposes, fishing, and water supply.

multistage remote sensing
a strategy for landscape Characterisation that involves gathering and analysing information at several geographic scales, ranging from generalised levels of detail at the national level through high levels of detail at the local scale.

municipal discharge
discharge of effluent from waste water treatment plants which receive waste water from households, commercial establishments, and industries in the coastal drainage basin. Combined sewer/separate storm overflows are included in this category.

municipal sewage
wastes (mostly liquid) originating from a community; may be composed of domestic wastewaters and/or industrial discharges.

municipal sludge
semi-liquid residue remaining from the treatment of municipal water and wastewater.

■ **municipal solid waste**
common garbage or trash generated by industries, businesses, institutions, and homes.

■ **mutagen/mutagenicity**
an agent that causes a permanent genetic change in a cell other than that which occurs during normal growth. Mutagenicity is the capacity of a chemical or physical agent to cause such permanent changes.

■ **mutualism**
relationship between two or more species that benefits all parties.

■ **mycorrhizal fungi**
a fungus living in a mutualistic association with plants and facilitating nutrient and water uptake.

■ **natural resources**
nutrients and minerals in the soil and deeper layers of the earth's crust; water; wild and domestic plants and animals; air; and other resources produced by the earth's natural processes.

■ **natural selection**
one of several gradual mechanisms through which evolution occurs. Process by which a particular beneficial gene (or set of genes) is reproduced more than other genes in succeeding generations due to selective pressures in the environment that favour the beneficial gene. The result of natural selection is a population that contains a greater proportion of organisms better adapted to certain environmental conditions.

■ **navigable waters**
traditionally, waters sufficiently deep and wide for navigation by all, or specified vessels.

■ **necrosis**
death of plant or animal cells or tissues. In plants, necrosis can discolour stems or leaves or kill a plant entirely.

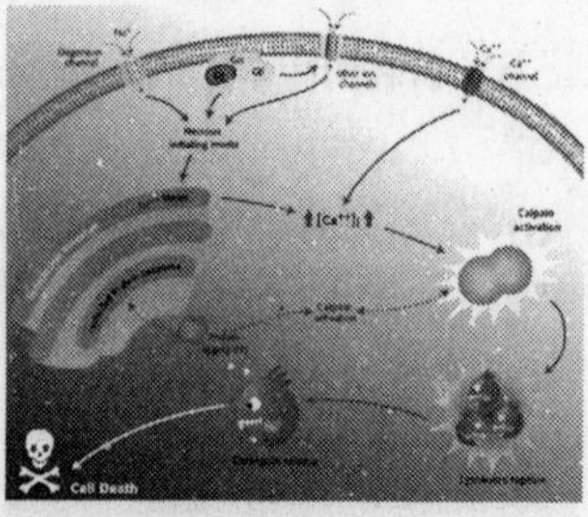

■ **nectarivore**
an animal that eats nectar.

■ **negative feedback loop**
situation in which a change in a certain direction provides information that causes a system to change less in that direction. This is a common regulatory mechanism and is widely used in animals to control hormone levels in the blood. For example, the hormones that control ovulation in humans are on a negative feedback loop.

■ **nekton**
animals, such as fish and whales, that move independently of water currents between the bottom and surface of the ocean.

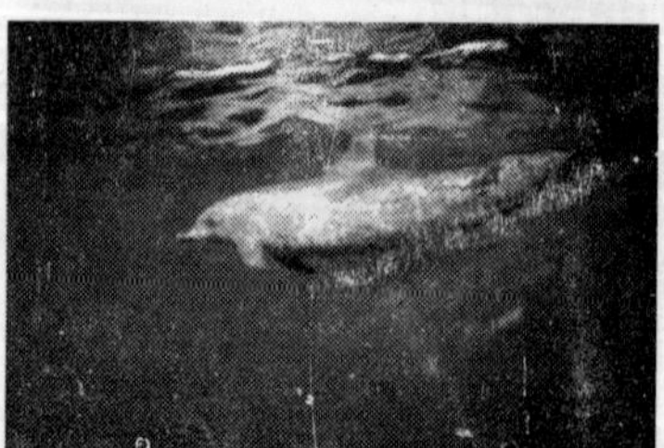

■ **nematocide**
a chemical agent which is destructive to nematodes.

■ **nephelometric**
method of measuring turbidity in a water sample by passing light through the sample and measuring the amount of the light that is deflected.

■ **netting**
a concept in which all emissions sources in the same area that owned or controlled by a single company are treated as one large source, thereby allowing flexibility in controlling individual sources in order to meet a single emissions standard. (See bubble.).

■ **neutral**
having a pH of 7 and thus being neither acidic nor basic.

■ **neutralisation**
decreasing the acidity or alkalinity of a substance by adding alkaline or acidic materials, respectively.

■ **new source**
any stationary source built or modified after publication of final or proposed regulations

that prescribe a given standard of performance.

■ **niche**

the location and role or job for which a species is well suited within its community, including its habitat, what it eats, its activities, and its interaction with other living things.

■ **nitrate**

a compound containing nitrogen that can exist in the atmosphere or as a dissolved gas in water and which can have harmful effects on humans and animals. Nitrates in water can cause severe illness in infants and domestic animals. A plant nutrient and inorganic fertiliser, nitrate is found in septic systems, animal feed lots, agricultural fertilisers, manure, industrial waste waters, sanitary landfills, and garbage dumps.

■ **nitric oxide (NO)**

a gas formed by combustion under high temperature and high pressure in an internal combustion engine; it is converted by sunlight and photochemical processes in ambient air to nitrogen oxide. NO is a precursor of ground-level ozone pollution, or smog..

■ **nitrification**

the process whereby ammonia in wastewater is oxidised to nitrite and then to nitrate by bacterial or chemical reactions.

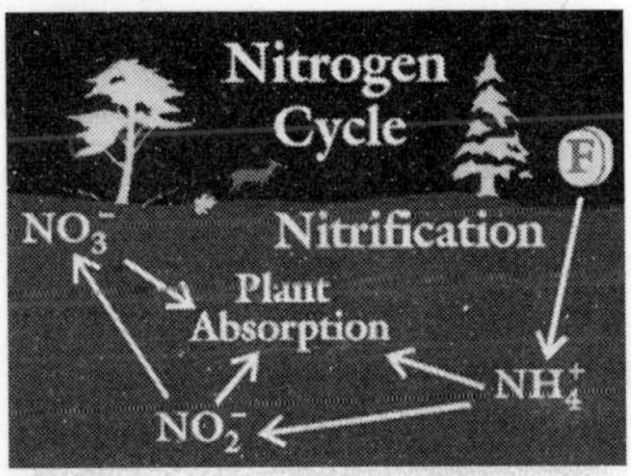

■ **nitrilotriacetic acid (NTA)**

a compound now replacing phosphates in detergents.

■ **nitrite**

1. an intermediate in the process of nitrification. 2. Nitrous oxide salts used in food preservation.

■ **nitrogen cycle**

cyclic movement of nitrogen in different chemical forms from the environment to organisms and then back to the environment.

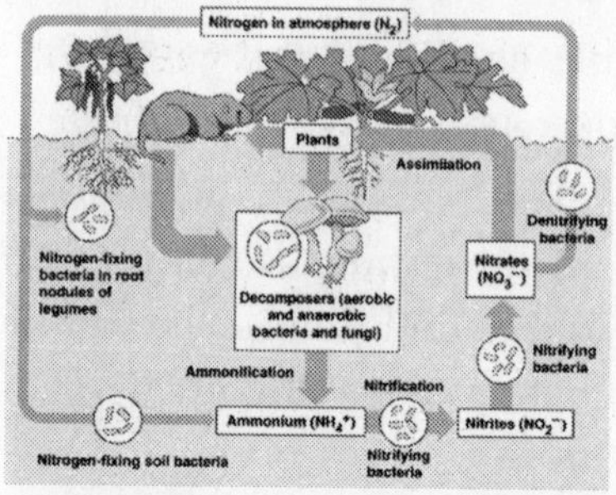

■ nitrogen dioxide (NO_2)
the result of nitric oxide combining with oxygen in the atmosphere; major component of photochemical smog.

■ nitrogen fixation
a process whereby *nitrogen fixing bacteria* living in mutualistic associations with plants convert atmospheric nitrogen to nitrogen compounds that plants can utilize directly.

■ nitrogen oxide (NO_x)
the result of photochemical reactions of nitric oxide in ambient air; major component of photochemical smog. Product of combustion from transportation and stationary sources and a major contributor to the formation of ozone in the troposphere and to acid deposition.

■ nitrogenous wastes
animal or vegetable residues that contain significant amounts of nitrogen.

■ nitrophenols
synthetic organopesticides containing carbon, hydrogen, nitrogen, and oxygen.

■ no observable adverse effect level
an exposure level at which there are no statistically or biologically significant increases in the frequency or severity of adverse effects between the exposed population and its appropriate control; some effects may be produced at this level, but they are not considered as adverse, or as precursors to adverse effects. In an experiment with several NOAELs, the regulatory focus is primarily on the highest one, leading to the common usage of the term NOAEL as the highest exposure without adverse effects.

■ no till
planting crops without prior seedbed preparation, into an existing cover crop, sod, or

crop residues, and eliminating subsequent tillage operations.

noble metal
chemically inactive metal such as gold; does not corrode easily.

nocturnal
active only at night.

node
where leaves grow from a plant stem.

noise
product-level or product-volume changes occurring during a test that are not related to a leak but may be mistaken for one.

non-aqueous phase liquid (NAPL)
contaminants that remain undiluted as the original bulk liquid in the subsurface, e.g. spilled oil.

nonbiodegradable
not able to be consumed and/or broken down by biological organisms. Non-biodegradable substances include plastics, aluminium, and many chemicals used in industry and agriculture. Particularly dangerous are non-biodegradable chemicals that are also toxic and tend to accumulate in organisms.

non-community water system
a public water system that is not a community water system; e.g. the water supply at a camp site or national park.

non-compliance coal
any coal that emits greater than 3.0 pounds of sulphur dioxide per million BTU when burned. Also known as high-sulphur coal.

non-contact cooling water
water used for cooling which does not come into direct contact with any raw material, product, byproduct, or waste.

non-conventional pollutant
any pollutant not statutorily listed or which is poorly understood by the scientific community.

non-degradation
an environmental policy which disallows any lowering of naturally occurring quality

regardless of pre-established health standards.

■ nondischarging treatment plant

a treatment plant that does not discharge treated wastewater into any stream or river. Most are pond systems that dispose of the total flow they receive by means of evaporation or percolation to groundwater, or facilities that dispose of their effluent by recycling or reuse (e.g. spray irrigation or groundwater discharge).

■ non-ferrous metals

nonmagnetic metals such as aluminium, lead, and copper. Products made all or in part from such metals include containers, packaging, appliances, furniture, electronic equipment and aluminium foil.

■ nonfriable asbestos-containing materials

any material containing more than one percent asbestos (as determined by Polarised Light Microscopy) that, when dry, cannot be crumbled, pulverised, or reduced to powder by hand pressure.

■ Non-Governmental Organisation (NGO)

a non-profit group or association organised outside of institutionalised political structures to realize particular social objectives (such as environmental protection) or serve particular constituencies (such as indigenous peoples). NGO activities range from research, information distribution, training, local organisation, and community service to legal advocacy, lobbying for legislative change, and civil disobedience. NGOs range in size from small groups within a particular community to huge membership groups with a national or international scope.

■ non-hazardous industrial waste

industrial process waste in wastewater not considered municipal solid waste or hazardous waste under RARA.

■ non-ionising electromagnetic radiation

1. radiation that does not change the structure of atoms but does heat tissue and may cause harmful biological effects.

2. Microwaves, radio waves, and low-frequency electromagnetic fields from high-voltage transmission lines.

■ **non-methane hydrocarbon (NMHC)**
the sum of all hydrocarbon air pollutants except methane; significant precursors to ozone formation.

■ **non-methane organic gases (NMOG)**
the sum of all organic air pollutants. Excluding methane; they account for aldehydes, ketones, alcohols, and other pollutants that are not hydrocarbons but are precursors of ozone.

■ **non-point sources**
diffuse pollution sources (i.e. without a single point of origin or not introduced into a receiving stream from a specific outlet). The pollutants are generally carried off the land by storm water. Common non-point sources are agriculture, forestry, urban, mining, construction, dams, channels, land disposal, saltwater intrusion, and city streets.

■ **non-potable**
water that is unsafe or unpalatable to drink because it contains pollutants, contaminants, minerals, or infective agents.

■ **non-renewable resource**
resource that exists in a fixed amount (stock) in various places in the earth's crust and has the potential for renewal only by geological, physical, and chemical processes taking place over hundreds of millions to billions of years. Examples are copper, aluminium, coal, and oil. We classify these resources as exhaustible because we are extracting and using them at a much faster rate than they were formed.

■ **non-road emissions**
pollutants emitted by combustion engines on farm and construction equipment, gasoline-powered lawn and garden equipment, and power boats and outboard motors.

■ **non-transient non-community water system**
a public water system that regularly serves at least 25 of the same non-resident persons

per day for more than six months per year.

■ no-observed-effect-level
exposure level at which there are no statistically or biological significant differences in the frequency or severity of any effect in the exposed or control populations.

■ northern hemisphere
the area of the earth above the equator.

■ northern temperate zone
the region between latitudes 23.5°N and 66.5°N.

■ nuclear energy
energy produced from changes in atomic nuclei.

■ nuclear reactors and support facilities
uranium mills, commercial power reactors, fuel reprocessing plants, and uranium enrichment facilities.

■ nuclear winter
prediction by some scientists that smoke and debris rising from massive fires of a nuclear war could block sunlight for weeks or months, cooling the earth's surface and producing climate changes that could, for example, negatively affect world agricultural and weather patterns.

■ nuclide
an atom characterised by the number of protons, neturons, and energy in the nucleus.

■ nutrient
any element or simple compound necessary for the health and survival of an organism. This includes air and water, as well as food.

■ nutrient cycling
all the processes by which nutrients are transferred from one organism to another. For instance, the carbon cycle includes uptake of carbon dioxide by plants, ingestion by animals, and respiration and decay of the animal.

■ nutrient pollution
contamination of water resources by excessive inputs of nutrients. In surface waters, excess algal production is a major concern.

ocean discharge waiver
a variance from Clean Water Act requirements for discharges into marine waters.

oceans
the largest bodies of water on earth.

odour threshold
the minimum odour of a water or air sample that can just be detected after successive dilutions with odourless water. Also called threshold odour.

office paper
high grade papers such as copier paper, computer printout, and stationary almost entirely made of uncoated chemical pulp, although some ground wood is used. Such waste is also generated in homes, schools, and elsewhere.

offsets
a concept whereby emissions from proposed new or modified stationary sources are balanced by reductions from existing sources to Stabilise total emissions. (See bubble, emissions trading, netting).

off-site facility
a hazardous waste treatment, storage or disposal area that is located away from the generating site.

offstream use
water withdrawn from surface or groundwater sources for use at another place.

oil and gas waste
gas and oil drilling muds, oil production brines, and other waste associated with exploration for, development and production of crude oil or natural gas.

oil desulphurisation
widely used precombustion method for reducing sulphur dioxide emissions from oil-burning power plants. The oil is treated with hydrogen, which removes some of the sulphur by forming hydrogen sulphide gas.

oil fingerprinting
a method that identifies sources of oil and allows spills to be traced to their source.

oil spill
an accidental or intentional discharge of oil which

reaches bodies of water. Can be controlled by chemical dispersion, combustion, mechanical containment, and/or adsorption. Spills from tanks and pipelines can also occur away from water bodies, contaminating the soil, getting into sewer systems and threatening underground water sources.

■ **old-growth forest**

virgin and old, second growth forests containing trees that are often hundreds, sometimes thousands, of years old. These are the richest forest biomes with the widest arrays of niche microhabitats and the broadest biodiversity, especially in the tropics.

■ **oligotrophic lakes**

deep clear lakes with few nutrients, little organic matter and a high dissolved-oxygen level.

■ **omnivore**

literally, an organism that will eat anything. Refers to animals who do not restrict their diet to just plants or other animals.

■ **onboard controls**

devices placed on vehicles to capture gasoline vapour during refuelling and route it to the engines when the vehicle is starting so that it can be efficiently burned.

■ **onconogenicity**

the capacity to induce cancer.

■ **one-hit model**

a mathematical model based on the biological theory that a single "hit" of some minimum critical amount of a carcinogen at a cellular target such as DNA can start an irreversible series events leading to a tumor.

■ on-site facility
a hazardous waste treatment, storage or disposal area that is located on the generating site.

■ opacity
the amount of light obscured by particulate pollution in the air; clear window glass has zero opacity, a brick wall is 100 percent opaque. Opacity is an indicator of changes in performance of particulate control systems.

■ open burning
uncontrolled fires in an open dump.

■ open dump
an uncovered site used for disposal of waste without environmental controls. (See dump.).

■ open system
a system, such as a living organism, in which both matter and energy are exchanged between the system and the environment.

■ operable unit
term for each of a number of separate activities undertaken as part of a Superfund site cleanup. A typical operable unit would be removal of drums and tanks from the surface of a site.

■ operating conditions
conditions specified in a RCRA permit that dictate how an incinerator must operate as it burns different waste types. A trial burn is used to identify operating conditions needed to meet specified performance standards.

■ operation and maintenance
1.activities conducted after a Superfund site action is completed to ensure that the action is effective. 2. Actions taken after construction to ensure that facilities constructed to treat waste water will be properly operated and maintained to achieve normative efficiency levels and prescribed effluent limitations in an optimum manner. 3. Ongoing asbestos management plan in a school or other public building, including regular inspections, various methods of maintaining asbestos in place, and removal when necessary.

optimal corrosion control treatment
an erosion control treatment that Minimises the lead and copper concentrations at users' taps while also ensuring that the treatment does not cause the water system to violate any national primary drinking water regulations.

optimum sustainable population
the number of animals which will result in the maximum productivity of the population or the species, keeping in mind the carrying capacity of the habitat and the health of the ecosystem.

oral toxicity
ability of a pesticide to cause injury when ingested.

organic
pertaining to compounds containing carbon. Also refers to living things or the materials made by living things. *inorganic* – ant.

organic chemicals/ compounds
naturally occurring (animal or plant-produced or synthetic) substances containing mainly carbon, hydrogen, nitrogen, and oxygen.

organic matter
carbonaceous waste contained in plant or animal matter and originating from domestic or industrial sources.

organism
all living things, including people, plants, animals, bacteria, and fungi.

organophosphates
pesticides that contain phosphorus; short-lived, but some can be toxic when first applied.

organophyllic
a substance that easily combines with organic compounds.

organotins
chemical compounds used in anti-foulant paints to protect the hulls of boats and ships, buoys, and pilings from marine organisms such as barnacles.

original generation point
where regulated medical or other material first becomes waste.

■ **orthodox seed**

seed that can be dried to moisture levels between 4 and 6 percent and kept at low temperatures.

■ **osmosis**

the passage of a liquid from a weak solution to a more concentrated solution across a semipermeable membrane that allows passage of the solvent (water) but not the dissolved solids.

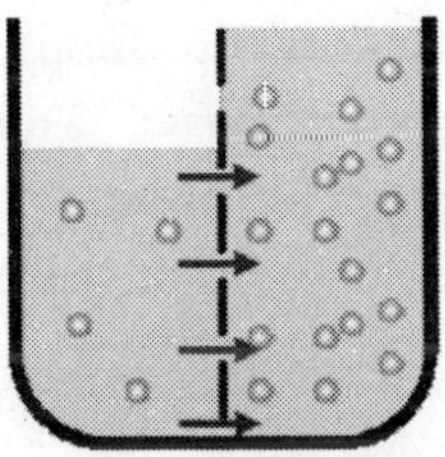

Osmosis
(Water moves by concentration gradient)

■ **other ferrous metals**

recyclable metals from strapping, furniture, and metal found in tires and consumer electronics but does not include metals found in construction materials or cars, locomotives, and ships.

■ **other glass**

recyclable glass from furniture, appliances, and consumer electronics. Does not include glass from transportation products (cars trucks or shipping containers) and construction or demolition debris. (See glass.).

■ **other nonferrous metals**

recyclable nonferrous metals such as lead, copper, and zinc from appliances, consumer electronics, and nonpackaging aluminium products. Does not include nonferrous metals from industrial applications and construction and demolition debris. (See nonferrous metals.).

■ **other paper**

for Recyclable paper from books, third-class mail, commercial printing, paper towels, plates and cups; and other nonpackaging paper such as posters, photographic papers, cards and games, milk cartons, folding boxes, bags, wrapping paper, and paperboard. Does not include wrapping paper or shipping cartons.

■ **other plastics**

recyclable plastic from appliances, eating utensils, plates, containers, toys, and various kinds of equipment. Does not include heavy-duty plastics such as yielding materials.

■ other solid waste
recyclable nonhazardous solid wastes, other than municipal solid waste, covered under Subtitle D of RARA. (See solid waste.).

■ other wood
recyclable wood from furniture, consumer electronics cabinets, and other nonpackaging wood products. Does not include lumber and tree stumps recovered from construction and demolition activities, and industrial process waste such as shavings and sawdust.

■ outdoor air supply
air brought into a building from outside.

■ outfall
the place where effluent is discharged into receiving waters.

■ overburden
layer of soil and rock overlying a mineral deposit, removed during surface mining.

■ overconsumption
situation in which some people consume much more than they need at the expense of those who can not meet their basic needs- and at the expense of earth's present and future life-support systems for humans and other forms of life.

■ overdraft
the pumping of water from a groundwater basin or aquifer in excess of the supply flowing into the basin; results in a depletion or "mining" of the groundwater in the basin. (See groundwater mining).

■ overfire air
air forced into the top of an incinerator or boiler to fan the flames.

■ overfishing
harvesting so many fish of a species (especially immature fish) that there is not enough breeding stock left to replenish the species, such that it is not profitable to harvest them, leading to commercial extinction .

■ overflow rate

one of the guidelines for design of the settling tanks and clarifers in a treatment plant; used by plant operators to determine if tanks and clarifiers are over or under-used.

■ overgrasing

destruction of vegetation when too many grasing animals feed too long and exceed the carrying capacity of a rangeland area.

■ overland flow

a land application technique that cleanses waste water by allowing it to flow over a sloped surface. As the water flows over the surface, contaminants are absorbed and the water is collected at the bottom of the slope for reuse.

■ overnutrition

diet so high in calories, saturated (animal) fats, salt, sugar, and processed foods, and so low in vegetables and fruits that the consumer runs high risks of diabetes, hypertension, heart disease, and other health hazards.

■ oversised regulated medical waste

medical waste that is too large for plastic bags or standard containers.

■ overturn

one complete cycle of top to bottom mixing of previously stratified water masses. This phenomenon may occur in spring or fall, or after storms, and results in uniformity of chemical and physical proper ties of water at all depths.

■ oxidant

a collective term for some of the primary constituents of photochemical smog.

■ oxidation

the chemical addition of oxygen to break down pollutants or organisac waste; e.g., destruction of chemicals such as cyanides, phenols, and organic sulphur compounds in sewage by bacterial and chemical means.

■ oxidation pond

a man-made (anthropogenic) body of water in which waste is consumed by bacteria, used most frequently with other waste-treatment processes; a sewage lagoon.

■ oxidation-reduction potential

the electric potential required to transfer electrons from one compound or element (the oxidant) to another compound (the reductant); used as a qualitative measure of the state of oxidation in water treatment systems.

■ oxygen

an atmospheric gas made up of two oxygen atoms that is necessary for respiration.

■ oxygen cycle

the recycling of oxygen-containing gases between plants and animals.

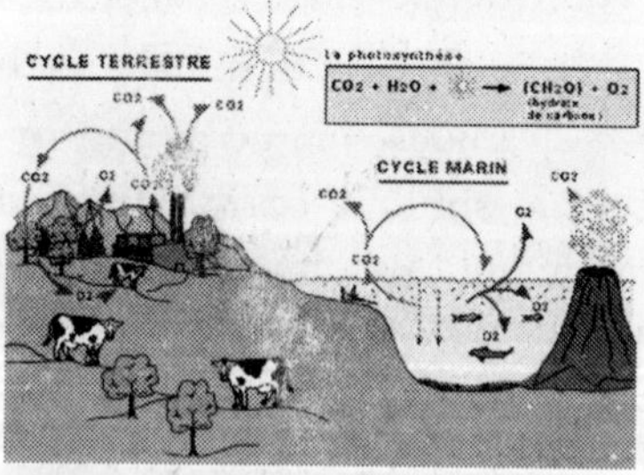

■ oxygenated fuels

gasoline which has been blended with alcohols or ethers that contain oxygen in order to reduce carbon monoxide and other emissions.

■ oxygenated solvent

an organic solvent containing oxygen as part of the molecular structure. Alcohols and ketones are oxygenated compounds often used as paint solvents.

■ ozonation/ozonator

application of ozone to water for disinfection or for taste and odour control. The ozonator is the device that does this.

■ ozone (O_3)

found in two layers of the atmosphere, the stratosphere and the troposphere. In the stratosphere (the atmospheric layer 7 to 10 miles or more above the earth's surface) ozone is a natural form of oxygen that provides a protective layer shielding the earth from ultraviolet radiation.In the troposphere (the layer extending up 7 to 10 miles from the earth's surface), ozone is a chemical oxidant and major component of photochemical smog. It can seriously impair the respiratory system. Ozone in the troposphere is produced through complex chemical reactions of nitrogen oxides, which are among the primary pollutants emitted by combustion sources; hydrocarbons, released into the atmosphere through the combustion, handling and processing of petroleum products; and sunlight.

■ ozone cycle

the ongoing process by which ozone breaks down and reforms in the ozone layer.

■ ozone depletion

destruction of the stratospheric ozone layer which shields the earth from ultraviolet radiation harmful to life. This destruction of ozone is caused by the breakdown of certain chlorine and/ or bromine containing compounds (chlorofluorocarbons or halons), which break down when they reach the stratosphere and then catalytically destroy ozone molecules.

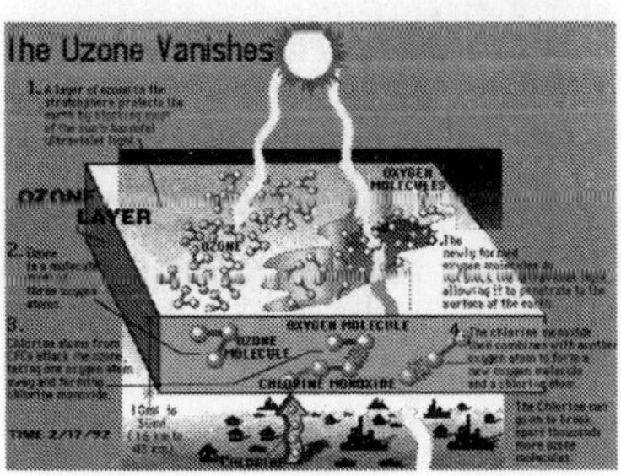

■ ozone hole

a thinning break in the stratospheric ozone layer. Designation of amount of such depletion as an "ozone hole" is made when the detected amount of depletion exceeds fifty percent. Seasonal ozone holes have been observed over both the Antarctic and Arctic regions, part of Canada, and the extreme northeastern United States.

■ ozone layer

the protective layer in the atmosphere, about 15 miles above the ground, that absorbs some of the sun's ultraviolet rays, thereby reducing the amount of potentially harmful radiation that reaches the earth's surface.

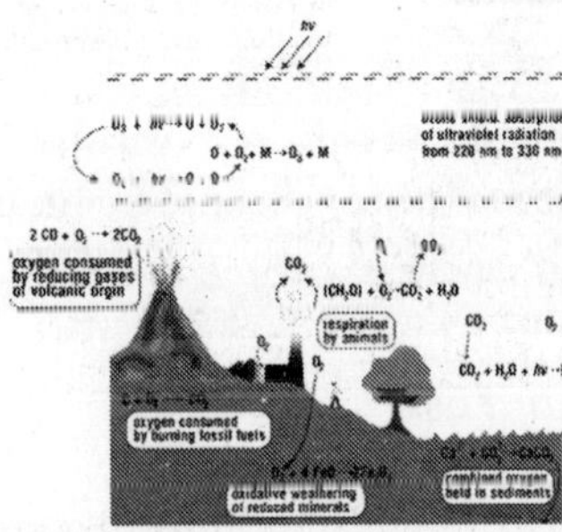

■ packaging

the assembly of one or more containers and any other components necessary to ensure minimum compliance with a program's storage and shipment packaging requirements. Also, the containers, etc. involved.

■ packed bed scrubber
an air pollution control device in which emissions pass through alkaline water to neutralise hydrogen chloride gas.

■ packed tower
a pollution control device that forces dirty air through a tower packed with crushed rock or wood chips while liquid is sprayed over the packing material. The pollutants in the air stream either dissolve or chemically react with the liquid.

■ packer
an inflatable gland, or balloon, used to create a temporary seal in a borehole, probe hole, well, or drive casing. It is made of rubber or non-reactive materials.

■ palatable water
water, at a desirable temperature, that is free from objectionable tastes, odours, colours, and turbidity.

■ paleoecology
the study of ancient ecosystems. Paleoecologists use data from such sources as tree rings, geologic deposits, fossils (pollen is a particularly popular tool), and coral bores to reconstruct the climate and ecology or ancient ecosystems.

■ pandemic
a widespread epidemic throughout an area, nation or the world.

■ paper
in the recycling business, refers to products and materials, including newspapers, magazines, office papers, corrugated containers, bags and some paperboard packaging that can be recycled into new paper products.

■ paper processor/plastics processor
intermediate facility where recovered paper or plastic products and materials are sorted, decontaminated, and prepared for final recycling.

■ parameter
a variable, measurable property whose value is a determinant of the characteristics of a system; e.g. temperature, pressure, and density are parameters of the atmosphere.

■ paraquat
a standard herbicide used to kill various types of crops, including marijuana. Causes lung damage if smoke from the crop is inhaled..

■ parasite
an organism that lives on or within a host (another organism); it obtains nutrients from the host without benefiting or killing (although it may damage) the host; parasitic, adj.; parasitism, n. a type of symbiotic relationship in which one organism benefits and the other does not.

■ paracitiom
a relationship in which one organism, a parasite, secures its nourishment by living on or inside a host organism at the expense of its host.

■ parataxonomists
field-trained biodiversity collection and inventory specialists recruited from local areas.

■ parshall flume
device used to measure the flow of water in an open channel.

■ participation rate
portion of population participating in a recycling program.

■ particle count
results of a microscopic examination of treated water with a special "particle counter" that classifies suspended particles by number and size.

■ particulate loading
the mass of particulates per unit volume of air or water.

■ particulates
1 fine liquid or solid particles such as dust, smoke, mist, fumes, or smog, found in air or emissions. 2. Very small solids suspended in water; they can vary in size, shape, density and electrical charge and can be gathered together by coagulation and flocculation.

■ partition coefficient
measure of the sorption phenomenon, whereby a pesticide is divided between the soil and water phase; also referred to as adsorption partition coefficient.

■ parts per billion (ppb)/ parts per million (ppm)
units commonly used to express contamination ratios, as in establishing the maximum permissible amount of a contaminant in water, land, or air.

■ passive smoking/second-hand smoke
inhalation of others' tobacco smoke.

■ passive treatment walls
technology in which a chemical reaction takes place when contaminated ground water comes in contact with a barrier such as limestone or a wall containing iron filings.

■ patent
a government grant of temporary monopoly rights on innovative processes or products.

■ pathogen
a disease-causing microorganism; a bacterium or virus.

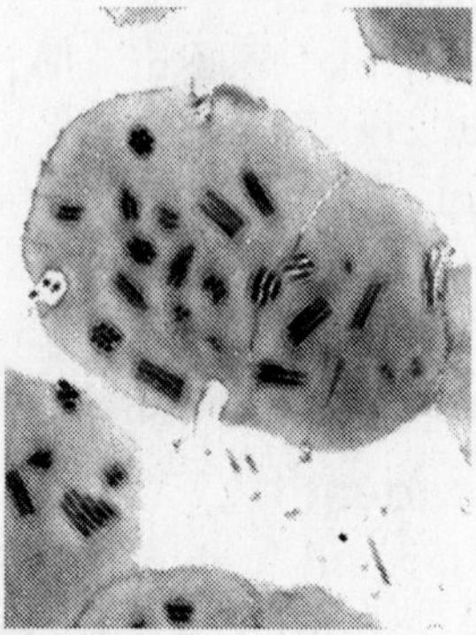

■ pathogenic
organism which causes a disease within another organism.

■ pathway
the physical course a chemical or pollutant takes from its source to the exposed organism.

■ peak levels
levels of airborne pollutant contaminants much higher than average or occurring for short periods of time in response to sudden releases.

■ pelagic
pelagic organisms swim through the ocean, and may rise to the surface, or sink to the bottom. They are not confined to live on the bottom as benthic organisms do.

■ percent saturatiuon
the amount of a substance that is dissolved in a solution compared to the amount that could be dissolved in it.

■ perched water
zone of unpressurised water held above the water table by impermeable rock or sediment.

■ percolating water
water that passes through rocks or soil under the force of gravity.

■ percolation
1.the movement of water downward and radially through subsurface soil layers, usually continuing downward to ground water. Can also involve upward movement of water.
2. slow seepage of water through a filter.

■ performance data (for incinerators)
information collected, during a trial burn, on concentrations of designated organic compounds and pollutants found in incinerator emissions.

■ periphyton
dense strands of algal growth that cover the water surface between the emergant aquatic plants. *Spirogyra* is commonly responsible for this growth.

■ permafrost
a layer of permanently frozen soil underground. An important feature of a tundra.

■ permeability
the rate at which liquids pass through soil or other materials in a specified direction.

■ permissible dose
the dose of a chemical that may be received by an individual without the expectation of a significantly harmful result.

■ persistence
refers to the length of time a compound stays in the environment, once introduced. A compound may persist for less than a second or indefinitely.

■ persistent pesticides
pesticides that do not break down chemically or break down very slowly and remain in the environment after a growing season.

personal measurement
a measurement collected from an individual's immediate environment.

personal protective equipment
clothing and equipment worn by pesticide mixers, loaders and applicators and re-entry workers, hazmat emergency responders, workers cleaning up Superfund sites, et. al., which is worn to reduce their exposure to potentially hazardous chemicals and other pollutants.

pest
an insect, rodent, nematode, fungus, weed or other form of terrestrial or aquatic plant or animal life that is injurious to health or the environment.

pest control operator
person or company that applies pesticides as a business (e.g. exterminator); usually describes household services, not agricultural applications.

pesticide
substances or mixture there of intended for preventing, destroying, repelling, or mitigating any pest. Also, any substance or mixture intended for use as a plant regulator, defoliant, or desiccant.

pesticide regulation notice
formal notice to pesticide registrants about important changes in regulatory policy, procedures, regulations.

PETE (Polyethylene Terepthalate)
thermoplastic material used in plastic soft drink and rigid containers.

petroleum
crude oil or any fraction thereof that is liquid under normal conditions of temperature and pressure. The term includes petroleum-based substances comprising a complex blend of hydrocarbons derived from crude oil through the process of separation, conversion, upgrading, and finishing, such as motor fuel, jet oil, lubricants, petroleum solvents, and used oil.

petroleum derivatives
chemicals formed when gasoline breaks down in contact with ground water.

pH
an expression of the intensity of the basic or acid condition of a liquid; may range from 0 to 14, where 0 is the most acid and 7 is neutral. Natural waters usually have a pH between 6.5 and 8.5.

pH scale
the scale, ranging from 0 to 14, used to measure the pH of a solution.

pharmacokinetics
the study of the way that drugs move through the body after they are swallowed or injected.

phenolphthalein alkalinity
the alkalinity in a water sample measured by the amount of standard acid needed to lower the pH to a level of 8.3 as indicated by the change of colour of the phenolphthalein from pink to clear.

phenols
organic compounds that are byproducts of petroleum refining, tanning, and textile, dye, and resin manufacturing. Low concentrations cause taste and odour problems in water; higher concentrations can kill aquatic life and humans.

phenotype
the morphological, physiological, biochemical, behavioral, and other properties of an organism that develop through the interaction of genes and environment.

phosphates
certain chemical compounds containing phosphorus.

phosphogypsum piles (stacks)
principal byproduct generated in production of phosphoric acid from phosphate rock. These piles may generate radioactive radon gas.

phosphorus
an essential chemical food element that can contribute to the eutrophication of lakes and other water bodies. Increased phosphorus levels result from discharge of phosphorus-containing materials into surface waters.

■ phosphorus cycle
cyclic movement of phosphorus, in varying chemical forms, from the environment to organisms and then back to the environment.

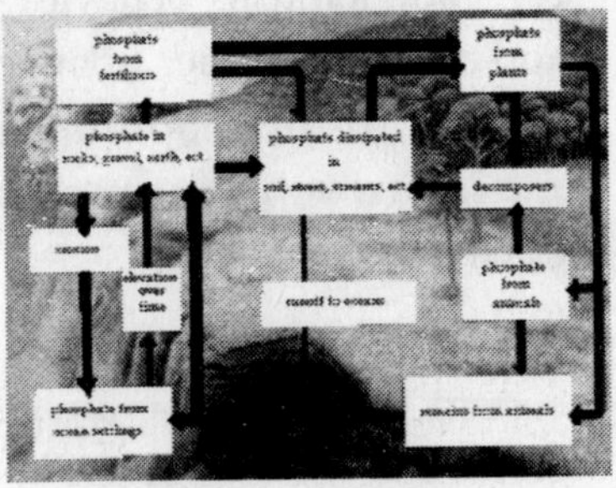

■ phosphorus plants
facilities using electric furnaces to produce elemental phosphorous for commercial use, such as high grade phosphoric acid, phosphate-based detergent, and organic chemicals use.

■ photic zone
region of the ocean through which light penetrates; and the place where photosynthetic marine organisms live.

■ photochemical oxidants
air pollutants formed by the action of sunlight on oxides of nitrogen and hydrocarbons.

■ photochemical smog
air pollution caused by chemical reactions of various pollutants emitted from different sources. (See photochemical oxidants.).

■ photosynthesis
the process by which plants use light energy trapped by chlorophyll to change carbon dioxide and water into food.

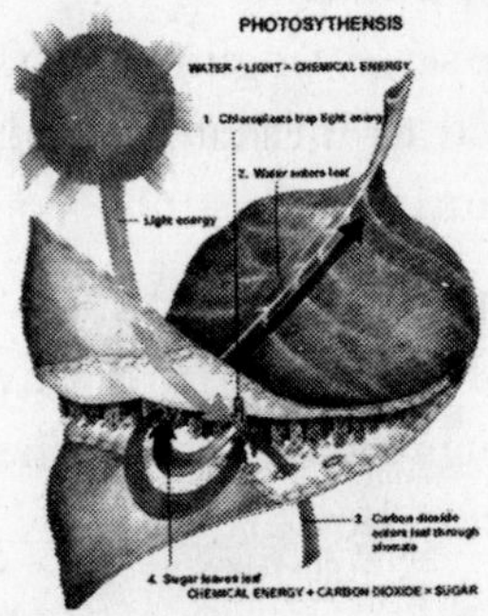

■ phylogenetic
pertaining to the evolutionary history of a particular group of organisms.

■ phylum
in taxonomy, a high-level category just beneath the kingdom and above the class; a group of related, similar classes.

■ physical and chemical treatment
processes generally used in large-scale wastewater treatment facilities. Physical processes may include air-stripping

or filtration. Chemical treatment includes coagulation, chlorination, or ozonation. The term can also refer to treatment of toxic materials in surface and ground waters, oil spills, and some methods of dealing with hazardous materials on or in the ground.

■ **phytoplankton**

tiny, free-floating, photosynthetic organisms in aquatic systems. They include diatoms, desmids, and dinoflagellates.

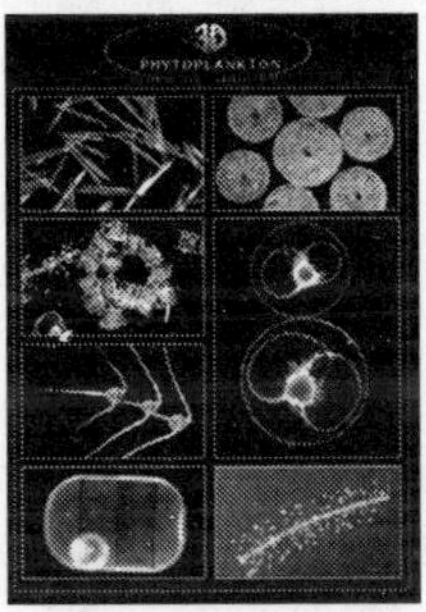

■ **phytoremediation**

low-cost remediation option for sites with widely dispersed contamination at low concentrations.

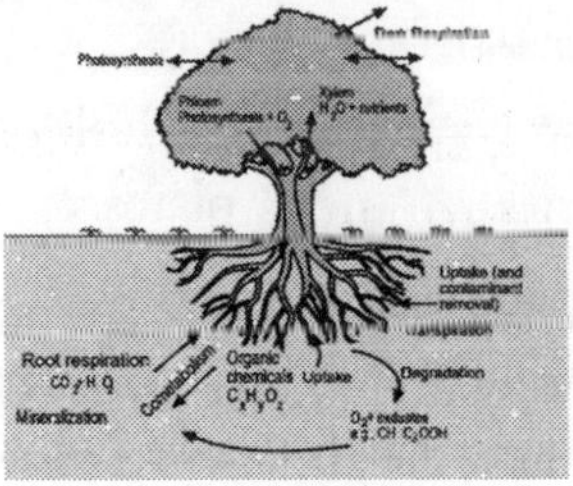

■ **phytotoxic**

harmful to plants.

■ **phytotreatment**

the cultivation of specialised plants that absorb specific contaminants from the soil through their roots or foliage. This reduces the concentration of contaminants in the soil, but incorporates them into biomasses that may be released back into the environment when the plant dies or is harvested.

■ **picocuries per liter (Pci/l)**

a unit of measure for levels of radon gas; becquerels per cubic meter is metric equivalent.

■ **piezometer**

a nonpumping well, generally of small diameter, for measuring the elevation of a water table.

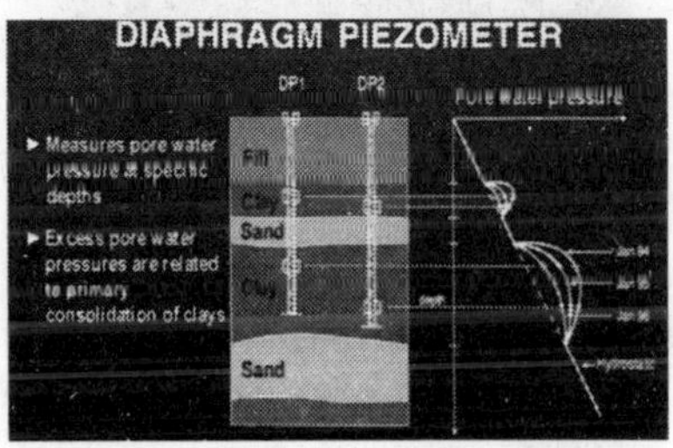

■ **pilot tests**

testing a cleanup technology under actual site conditions to identify potential problems prior to full-scale implementation.

■ pioneer species

first hardy, often xerophytic , species (often microbes, mosses, and lichens) that begin colonizing a site as the first stage of ecological succession.

■ plankton

very small, free-floating organisms of the ocean or other aquatic systems, including phytoplankton, which produce their own nutrients through photosynthesis, or zooplankton, which get their nutrients from organisms.

Plankton

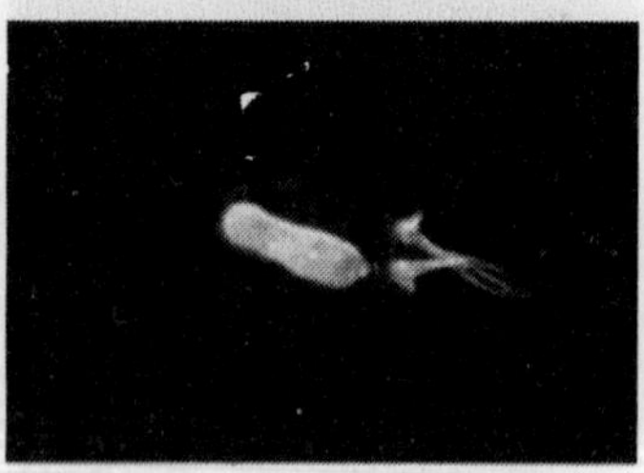

■ plasma-arc reactor

an incinerator that operates at extremely high temperatures; treats highly toxic wastes that do not burn easily.

■ plasmid

a circular piece of DNA that exists apart from the chromosome and replicates independently of it. Bacterial plasmids carry information that renders the bacteria resistant to antibiotics. Plasmids are often used in genetic engineering to carry desired genes into organisms.

Plasmid Insertion

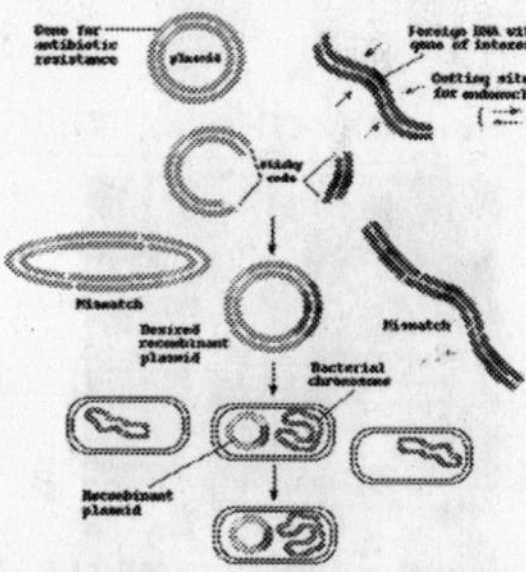

■ plastics

non-metallic chemoreactive compounds Moulded into rigid or pliable construction materials, fabrics, etc.

■ plate tower scrubber

an air pollution control device that neutralises hydrogen chloride gas by bubbling alkaline water through holes in a series of metal plates.

■ plug flow

type of flow the occurs in tanks, basins, or reactors

when a slug of water moves through without ever dispersing or mixing with the rest of the water flowing through.

plugging

act or process of stopping the flow of water, oil, or gas into or out of a formation through a borehole or well penetrating that formation.

plume

1. a visible or measurable discharge of a contaminant from a given point of origin. Can be visible or thermal in water, or visible in the air as, for example, a plume of smoke. 2 The area of radiation leaking from a damaged reactor. 3. Area downwind within which a release could be dangerous for those exposed to leaking fumes.

plutonium

a radioactive metallic element chemically similar to uranium.

PM-10/PM-2.5

PM 10 is measure of particles in the atmosphere with a diameter of less than ten or equal to a nominal 10 micrometers. PM-2.5 is a measure of smaller particles in the air.

pneumoconiosis

health conditions characterised by permanent deposition of substantial amounts of particulate matter in the lungs and by the tissue reaction to its presence; can range from relatively harmless forms of sclerosis to the destructive fibrotic effect of silicosis.

point source

a stationary location or fixed facility from which pollutants are discharged; any single identifiable source of pollution; e.g. a pipe, ditch, ship, ore pit, factory smokestack.

point-of-contact measurement of exposure

estimating exposure by measuring concentrations over

time (while the exposure is taking place) at or near the place where it is occurring.

■ point-of-disinfectant application

the point where disinfectant is applied and water downstream of that point is not subject to recontamination by surface water runoff.

■ point-of-use treatment device

treatment device applied to a single tap to reduce contaminants in the drinking water at the one faucet.

■ pollen

the fertilising element of flowering plants; background air pollutant.

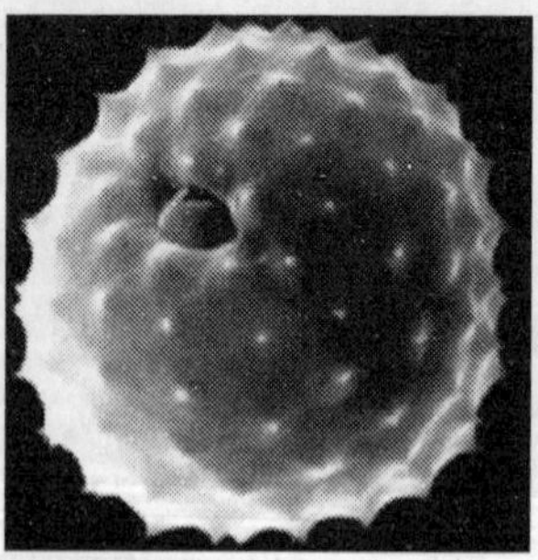

■ pollinator

animal which carries pollen from one seed plant to another, unwittingly aiding the plant in its reproduction. Common pollinators include insects, especially bees, butterflies, and moths, birds, and bats.

■ pollutant

generally, any substance introduced into the environment that adversely affects the usefulness of a resource or the health of humans, animals, or ecosystems..

■ pollutant pathways

avenues for distribution of pollutants. In most buildings, for example, HVAC systems are the primary pathways although all building components can interact to affect how air movement distributes pollutants.

■ pollutant standard index

indicator of one or more pollutants that may be used to inform the public about the potential for adverse health effects from air pollution in major cities.

■ pollution

generally, the presence of a substance in the environment that because of its chemical composition or quantity prevents the functioning of natural processes and produces undesirable environmental and health effects.

■ **polonium**
a radioactive element that occurs in pitchblende and other uranium-containing ores.

■ **polyelectrolytes**
synthetic chemicals that help solids to clump during sewage treatment.

■ **polymer**
a natural or synthetic chemical structure where two or more like molecules are joined to form a more complex molecular structure (e.g. polyethylene in plastic).

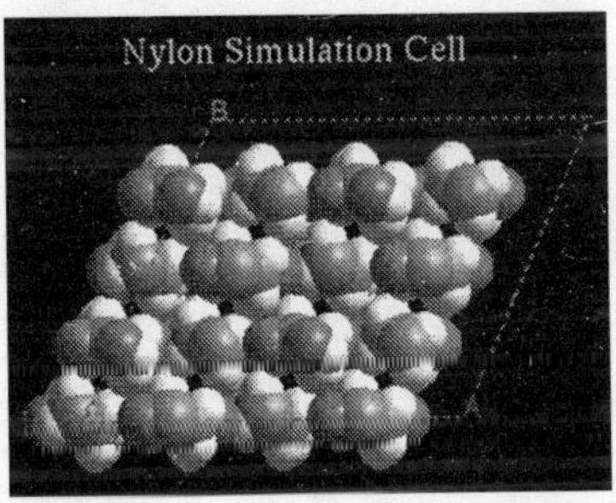

■ **polyp**
a tiny, tubelike marine animal of which live coral is made, one end of which is attached to the sea bottom, to rocks, or to one another and the opposite end of which is a mouth surrounded by finger-like, stinging tentacles.

■ **polyvinyl chloride (PVC)**
a tough, environmentally indestructible plastic that releases hydrochloric acid when burned.

■ **population**
a group of interbreeding organisms occupying a particular space; the number of humans or other living creatures in a designated area.

■ **population at risk**
a population subgroup that is more likely to be exposed to a chemical, or is more sensitive to the chemical, than is the general population.

■ **population change**
an increase or decrease in the size of a population. It is equal to (births + immigration) - (deaths + emigration).

■ **population density**
number of organisms in a particular population found in a specified area.

■ **population dispersion**
general pattern in which the members of a population are

arranged throughout its habitat.

■ population distribution

variation of population density over a particular geographical area. For example, a country has a high population density in its urban areas and a much lower population density in rural areas.

■ porosity

degree to which soil, gravel, sediment, or rock is permeated with pores or cavities through which water or air can move.

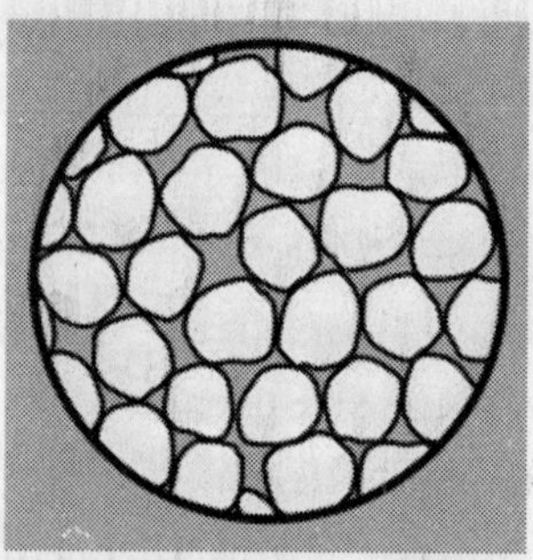

■ portal-of-entry effect

a local effect produced in the tissue or organ of first contact between a toxicant and the biological system.

■ positive feedback loop

situation in which a change in a certain direction provides information that causes a system to change further in the same direction. This can lead to a runaway or vicious cycle.

■ post-chlorination

addition of chlorine to plant effluent for disinfectant purposes after the effluent has been treated.

■ post-closure

the time period following the shutdown of a waste management or manufacturing facility; for monitoring purposes, often considered to be 30 years.

■ post-consumer materials/waste

recovered materials that are diverted from municipal solid waste for the purpose of collection, recycling, and disposition.

■ post-consumer recycling

use of materials generated from residential and consumer waste for new or similar purposes; e.g. converting wastepaper from offices into corrugated boxes or newsprint.

■ potable water

water that is safe for drinking and cooking.

■ **potential dose**
the amount of a compound contained in material swallowed, breathed, or applied to the skin.

■ **potentially renewable resource**
resource that theoretically can last indefinitely without reducing the available supply, either because it is replaced more rapidly through natural processes than are nonrenewable resources or because it is potentially inexhaustible (solar energy). Examples are trees in forests, grasses in grasslands, wild animals, fresh surface water in lakes and streams, most groundwater, fresh air, and fertile soil. If such a resource is used faster than it is replenished, it can be depleted and converted into a nonrenewable resource.

■ **potentiation**
the ability of one chemical to increase the effect of another chemical.

■ **potentiometric surface**
the surface to which water in an aquifer can rise by hydrostatic pressure.

■ **poverty**
inability to meet basic needs for food, clothing, and shelter.

■ **precautionary principle**
when information about potential risks is incomplete, basing decisions about the best ways to manage or reduce risks on a preference for avoiding unnecessary health risks instead of on unnecessary economic expenditures.

■ **prechlorination**
the addition of chlorine at the headworks of a treatment plant prior to other treatment processes. Done mainly for disinfection and control of tastes, odours, and aquatic growths, and to aid in coagulation and settling,

■ **precipitate**
a substance separated from a solution or suspension by chemical or physical change.

■ **precipitation**
removal of hazardous solids from liquid waste to permit safe disposal; removal of particles from airborne emissions as in rain (e.g. acid precipitation).

precipitator
pollution control device that collects particles from an air stream.

pre-consumer materials/waste
materials generated in manufacturing and converting processes such as manufacturing scrap and trimmings and cuttings. Includes print overruns, overissue publications, and obsolete inventories.

precursor
in photochemistry, a compound antecedent to a pollutant. For example, volatile organic compounds (VOCs) and nitric oxides of nitrogen react in sunlight to form ozone or other photochemical oxidants. As such, VOCs and oxides of nitrogen are precursors.

predator
organism which hunts and eats other organisms. This includes both carnivores, which eat animals, and herbivores, which eat plants.

predator control
a predator-prey interaction in which the predator controls the prey population size; that is, in which the predator population is the limiting factor for the prey population size.

pre-harvest interval
the time between the last pesticide application and harvest of the treated crops.

preliminary assessment
the process of collecting and reviewing available information about a known or suspected waste site or release.

prescriptive
water rights which are acquired by diverting water and putting it to use in accordance with specified procedures; e.g. filing a request with a state agency to use unused water in a stream, river, or lake.

pressed wood products
materials used in building and furniture construction that are made from wood veneers, particles, or fibers bonded together with an adhesive under heat and pressure.

pressure sewers
a system of pipes in which water, wastewater, or other

liquid is pumped to a higher elevation.

■ pressure, static
in flowing air, the total pressure minus velocity pressure, pushing equally in all directions.

■ pressure, total
in flowing air, the sum of the static and velocity pressures.

■ pressure, velocity
in flowing air, the pressure due to velocity and density of air.

■ pretreatment
processes used to reduce, eliminate, or alter the nature of wastewater pollutants from non-domestic sources before they are discharged into publicly owned treatment works (POTWs).

■ prevailing winds
winds that blow consistently from one direction.

■ prevalent level samples
air samples taken under normal conditions (also known as ambient background samples).

■ prevalent levels
levels of airborne contaminant occurring under normal conditions.

■ prey
organism hunted and eaten by a predator.

■ primacy
having the primary responsibility for administering and enforcing regulations.

■ primary effect
an effect where the stressor acts directly on the ecological component of interest, not on other parts of the ecosystem. (See secondary effect.).

■ primary producer
an organism, such as a plant or microbe, that makes its own food and forms the bottom-most tier in a trophic system. Primary producers are the basis of the food web in most ecosystems (the exceptions are open system communities based entirely on scavenging nutrients flushed into the system from elsewhere, such as some deep sea communities—though even in these cases,

the food flushed into the system comes from another system where primary producers are the basis of the trophic pyramid). Primary producers are able to convert abiotic raw materials into biotic tissue, either by capturing the sun's energy through photosynthesis (plants) or by harnessing the energy in chemical bonds through chemosynthesis (some microbes).

primary productivity
the transformation of chemical or solar energy to biomass. Most primary production occurs through photosynthesis, whereby green plants convert solar energy, carbon dioxide, and water to glucose and eventually to plant tissue. In addition, some bacteria in the deep sea can convert chemical energy to biomass through chemosynthesis. Primary production refers to the *amount* of material produced. *Net primary production* is the measure of the actual accumulation of biomass after some of the products of photosynthesis (or chemosynthesis) are expended for the plant's own maintenance. Productivity, or the *rate* of production, is affected by various environmental factors, including the amount of solar radiation, the availability of water and mineral nutrients, and temperature. *(See "Biodiversity in Ecosystems").*

primary standards
national ambient air quality standards designed to protect human health with an adequate margin for safety. (See National Ambient Air Quality Standards, secondary standards.).

primary waste treatment
first steps in wastewater treatment; screens and sedimentation tanks are used to remove most materials that float or will settle. Primary treatment removes about 30 percent of carbonaceous biochemical oxygen demand from domestic sewage.

principal organic hazardous constituents
hazardous compounds monitored during an incinerator's

trial burn, selected for high concentration in the waste feed and difficulty of combustion.

■ **prior appropriation**
a doctrine of water law that allocates the rights to use water on a first-come, first-served basis.

■ **probability of detection**
the likelihood, expressed as a percentage, that a test method will correctly identify a leaking tank.

■ **process variable**
a physical or chemical quantity which is usually measured and controlled in the operation of a water treatment plant or industrial plant.

■ **process verification**
verifying that process raw materials, water usage, waste treatment processes, production rate and other facts relative to quantity and quality of pollutants contained in discharges are substantially described in the permit application and the issued permit.

■ **process wastewater**
any water that comes into contact with any raw material, product, byproduct, or waste.

■ **process weight**
total weight of all materials, including fuel, used in a manufacturing process; used to calculate the allowable particulate emission rate.

■ **producer**
any organism which brings energy into an ecosystem from inorganic sources. Most plants and many protists are producers.

■ **product level**
the level of a product in a storage tank.

■ **product water**
water that has passed through a water treatment plant and is ready to be delivered to consumers.

■ **products of incomplete combustion (PICs)**
organic compounds formed by combustion. Usually generated in small amounts and sometimes toxic, PICs are heat-altered versions of the original material fed into the

incinerator (e.g. charcoal is a P.I.C. from burning wood).

■ **propellant**
liquid in a self-pressurised pesticide product that expels the active ingredient from its container.

■ **proportionate mortality ratio**
the number of deaths from a specific cause in a specific period of time per 100 deaths from all causes in the same time period.

■ **proposed plan**
a plan for a site cleanup that is available to the public for comment.

■ **proteins**
complex nitrogenous organic compounds of high molecular weight made of amino acids; essential for growth and repair of animal tissue. Many, but not all, proteins are enzymes.

■ **protocol**
a series of formal steps for conducting a test.

■ **protoplast**
a membrane-bound cell from which the outer wall has been partially or completely removed. The term often is applied to plant cells.

■ **protozoa**
one-celled animals that are larger and more complex than bacteria. May cause disease.

■ **public health approach**
regulatory and voluntary focus on effective and feasible risk management actions at the national and community level to reduce human exposures and risks, with priority given to reducing exposures with the biggest impacts in terms of the number affected and severity of effect.

■ **public health context**
the incidence, prevalence, and severity of diseases in communities or populations and the factors that account for them, including infections, exposure to pollutants, and other exposures or activities.

■ **public water system**
a system that provides piped water for human consumption to at least 15 service connec-

tions or regularly serves 25 individuals.

■ publicly owned treatment works

a waste-treatment works owned by a state, unit of local government, or Indian tribe, usually designed to treat domestic wastewaters.

■ pumping station

mechanical device installed in sewer or water system or other liquid-carrying pipelines to move the liquids to a higher level.

■ pumping test

a test conducted to determine aquifer or well characteristics.

■ purging

removing stagnant air or water from sampling zone or equipment prior to sample collection.

■ putrefaction

biological decomposition of organic matter; associated with anaerobic conditions.

■ putrescible

able to rot quickly enough to cause odours and attract flies.

■ pyramid of biomass

diagram representing the biomass (total dry weight of living organisms) that can be supported at each trophic level in a food web. The bottom of the pyramid is comprised of primary producers , while the peak of the pyramid is topped by one (or at most a small handful) apex predator . Humans are abnormal in that we cross all ecosystems and biomass pyramids, and in almost every one (excepting the polar caps and deepest of oceanic environments) we are the dominant apex predator.

■ pyramid of energy flow

also called a trophic pyramid. Diagram representing the flow of energy through each trophic level in a food chain or food web. With each energy transfer, only a small part (typically 10%) of the usable energy entering one trophic level is transferred to the organisms at the next trophic level, with the remaining 90% lost as heat or expended in metabolic processes.

pyrolysis
decomposition of a chemical by extreme heat.

qualitative use assessment
report summarising the major uses of a pesticide including percentage of crop treated, and amount of pesticide used on a site.

quench tank
a water-filled tank used to cool incinerator residues or hot materials during industrial processes.

radiation
transmission of energy though space or any medium. Also known as radiant energy.

radiation standards
regulations that set maximum exposure limits for protection of the public from radioactive materials.

radioactive decay
spontaneous change in an atom by emission of of charged particles and/or gamma rays; also known as radioactive disintegration and radioactivity.

radioactive substances
substances that emit Ionising radiation.

radioisotopes
chemical variants of radioactive elements with potentially oncogenic, teratogenic, and mutagenic effects on the human body.

radionuclide
radioactive particle, man-made (anthropogenic) or natural, with a distinct atomic weight number. Can have a long life as soil or water pollutant.

radius of influence
1. the radial distance from the center of a wellbore to the point where there is no lowering of the water table or potentiometric surface (the edge of the cone of depression);
2. the radial distance from an extraction well that has adequate air flow for effective removal of contaminants when a vacuum is applied to the extraction well.

radius of vulnerability zone
the maximum distance from the point of release of a hazardous

substance in which the airborne concentration could reach the level of concern under specified weather conditions.

■ radon

a colourless naturally occurring, radioactive, inert gas formed by radioactive decay of radium atoms in soil or rocks.

■ radon daughters/radon progeny

short-lived radioactive decay products of radon that decay into longer-lived lead isotopes that can attach themselves to airborne dust and other particles and, if inhaled, damage the linings of the lungs.

■ radon decay products

a term used to refer collectively to the immediate products of the radon decay chain. These include Po-218, Pb-214, Bi-214, and Po-214, which have an average combined half life of about 30 minutes.

■ rainbow report

comprehensive document giving the status of all pesticides now or ever in registration or special reviews. Known as the "rainbow report" because chapters are printed on different colours of paper.

■ rasp

a machine that grinds waste into a manageable material and helps prevent odour.

■ raw agricultural commodity

an unprocessed human food or animal feed crop (e.g., raw carrots, apples, corn, or eggs.).

■ raw sewage

untreated wastewater and its contents.

■ raw water

intake water prior to any treatment or use.

■ reaeration

introduction of air into the lower layers of a reservoir. As the air bubbles form and rise through the water, the oxygen dissolves into the water and replenishes the dissolved oxygen. The rising bubbles also cause the lower waters to rise to the surface where they take on oxygen from the atmosphere.

■ real-time monitoring

monitoring and measuring environmental developments

with technology and communications systems that provide time-relevant information to the public in an easily understood format people can use in day-to-day decision-making about their health and the environment.

reasonable maximum exposure
the maximum exposure reasonably expected to occur in a population.

reasonable worst case
an estimate of the individual dose, exposure, or risk level received by an individual in a defined population that is greater than the 90th percentile but less than that received by anyone in the 98th percentile in the same population.

reasonably available control measures (racm)
a broadly defined term referring to technological and other measures for pollution control.

reasonably available control technology (RACT)
control technology that is reasonably available, and both technologically and economically feasible. Usually applied to existing sources in nonattainment areas; in most cases is less stringent than new source performance standards.

recalcitrant seed
seed that does not survive drying and freezing.

recarbonisation
process in which carbon dioxide is bubbled into water being treated to lower the pH.

receiving waters
a river, lake, ocean, stream or other watercourse into which wastewater or treated effluent is discharged.

receptor
ecological entity exposed to a stressor.

recharge
the process by which water is added to a zone of saturation, usually by percolation from the soil surface; e.g., the recharge of an aquifer.

recharge area
a land area in which water reaches the zone of saturation from surface infiltration, e.g.,

where rainwater soaks through the earth to reach an aquifer.

■ recharge rate

the quantity of water per unit of time that replenishes or refills an aquifer.

■ reclamation

(In recycling) restoration of materials found in the waste stream to a beneficial use which may be for purposes other than the original use.

■ recombinant bacteria

a microorganism whose genetic makeup has been altered by deliberate introduction of new genetic elements. The offspring of these altered bacteria also contain these new genetic elements; i.e. they "breed true."

■ recombinant DNA

the new DNA that is formed by combining pieces of DNA from different organisms or cells.

■ recommended maximum contaminant level

the maximum level of a contaminant in drinking water at which no known or anticipated adverse effect on human health would occur, and that includes an adequate margin of safety. Recommended levels are nonenforceable health goals. (See maximum contaminant level.).

■ reconstructed source

facility in which components are replaced to such an extent that the fixed capital cost of the new components exceeds 50 percent of the capital cost of constructing a comparable brand-new facility. New-source performance standards may be applied to sources reconstructed after the proposal of the standard if it is technologically and economically feasible to meet the standards.

■ reconstruction of dose

estimating exposure after it has occurred by using evidence within an organism such as chemical levels in tissue or fluids.

■ recovery rate

percentage of usable recycled materials that have been removed from the total amount of municipal solid waste generated in a specific area or by a specific business.

recycle/reuse
minimising waste generation by recovering and reprocessing usable products that might otherwise become waste (.i.e. recycling of aluminium cans, paper, and bottles, etc.).

recycling and reuse business assistance centers
located in state solid-waste or economic-development agencies, these centers provide recycling businesses with customised and targeted assistance.

recycling economic development advocates
individuals hired by state or tribal economic development offices to focus financial, marketing, and permitting resources on creating recycling businesses.

recycling mill
facility where recovered materials are remanufactured into new products.

recycling technical assistance partnership national network
a national information-sharing resource designed to help businesses and manufacturers increase their use of recovered materials.

red bag waste
see **infectious waste.**

red tide
a proliferation of a marine plankton toxic and often fatal to fish, perhaps stimulated by the addition of nutrients. A tide can be red, green, or brown, depending on the colouration of the plankton.

redemption program
program in which consumers are monetarily compensated for the collection of recyclable materials, generally through prepaid deposits or taxes on beverage containers. In some states or localities legislation has enacted re-

demption programs to help prevent roadside litter.

■ **reduce**
to use less.

■ **reduction**
the addition of hydrogen, removal of oxygen, or addition of electrons to an element or compound.

■ **re-entry**
(In indoor air program) Refers to air exhausted from a building that is immediately brought back into the system through the air intake and other openings.

■ **reentry interval**
the period of time immediately following the application of a pesticide during which unprotected workers should not enter a field.

■ **reference dose (RfD)**
the RfD is a numerical estimate of a daily oral exposure to the human population, including sensitive subgroups such as children, that is not likely to cause harmful effects during a lifetime. RfDs are generally used for health effects that are thought to have a threshold or low dose limit for producing effects.

■ **reformulated gasoline**
gasoline with a different composition from conventional gasoline (e.g., lower aromatics content) that cuts air pollutants.

■ **refueling emissions**
emissions released during vehicle re-fueling.

■ **refuse**
see **solid waste.**

■ **refuse reclamation**
conversion of solid waste into useful products; e.g., composting organic wastes to make soil conditioners or separating aluminium and other metals for recycling.

■ **regeneration**
manipulation of cells to cause them to develop into whole plants.

■ **registrant**
any manufacturer or formulator who obtains registration for a pesticide active ingredient or product.

■ **registration standards**
published documents which include summary reviews of

the data available on a pesticide's active ingredient, data gaps, and the Agency's existing regulatory position on the pesticide.

■ regulated asbestos-containing material
friable asbestos material or nonfriable ACM that will be or has been subjected to sanding, grinding, cutting, or abrading or has crumbled, or been pulverised or reduced to powder in the course of demolition or renovation operations.

■ rehabilitation
the recovery of specific ecosystem services in a degraded ecosystem or habitat.

■ relative ecological sustainability
ability of an ecosystem to maintain relative ecological integrity indefinitely.

■ relative permeability
the permeability of a rock to gas, NAIL, or water, when any two or more are present.

■ relative risk assessment
estimating the risks associated with different stressors or management actions.

■ release
any spilling, leaking, pumping, pouring, emitting, emptying, discharging, injecting, escaping, leaching, dumping, or disposing into the environment of a hazardous or toxic chemical or extremely hazardous substance.

■ remedial action
the actual construction or implementation phase of a Superfund site cleanup that follows remedial design.

■ remedial design
a phase of remedial action that follows the remedial investigation/feasibility study and includes development of engineering drawings and specifications for a site cleanup.

■ remedial investigation
an in-depth study designed to gather data needed to determine the nature and extent of contamination at a Superfund site; establish site cleanup criteria; identify preliminary alternatives for remedial action; and support technical and cost analyses of alternatives. The remedial investigation is usu-

ally done with the feasibility study. Together they are usually referred to as the "RI/FS".

■ remedial response

long-term action that stops or substantially reduces a release or threat of a release of hazardous substances that is serious but not an immediate threat to public health.

■ remediation

1. cleanup or other methods used to remove or contain a toxic spill or hazardous materials from a Superfund site; 2. for the Asbestos Hazard Emergency Response program, abatement methods including evaluation, repair, enclosure, encapsulation, or removal of greater than 3 linear feet or square feet of asbestos-containing materials from a building.

■ remote sensing

the collection and interpretation of information about an object without physical contact with the object; e.g., satellite imaging, aerial photography, and open path measurements.

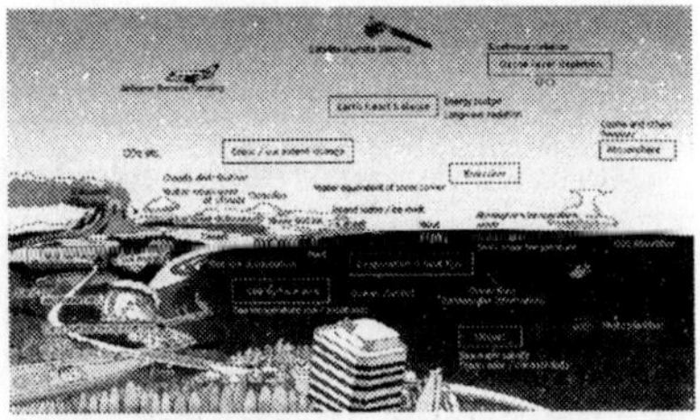

■ removal action

short-term immediate actions taken to address releases of hazardous substances that require expedited response. (See cleanup.).

■ repair

to fix.

■ repeat compliance period

any subsequent compliance period after the initial one.

■ repowering

rebuilding and replacing major components of a power plant instead of building a new one.

■ representative sample

a portion of material or water that is as nearly identical in content and consistency as possible to that in the larger body of material or water being sampled.

reserve capacity
extra treatment capacity built into solid waste and wastewater treatment plants and interceptor sewers to accommodate flow increases due to future population growth.

reservoir
any natural or artificial holding area used to store, regulate, or control water.

residential use
pesticide application in and around houses, office buildings, apartment buildings, motels, and other living or working areas.

residential waste
waste generated in single and multi-family homes, including newspapers, clothing, disposable tableware, food packaging, cans, bottles, food scraps, and yard trimmings other than those that are diverted to backyard composting. (See Household hazardous waste.).

residual
amount of a pollutant remaining in the environment after a natural or technological process has taken place; e.g., the sludge remaining after initial wastewater treatment, or particulates remaining in air after it passes through a scrubbing or other process.

residual risk
the extent of health risk from air pollutants remaining after application of the Maximum Achievable Control Technology (MACT).

residual saturation
saturation level below which fluid drainage will not occur.

residue
the dry solids remaining after the evaporation of a sample of water or sludge.

resistance
for plants and animals, the ability to withstand poor environmental conditions or attacks by chemicals or disease. May be inborn or acquired.

resource economics
the study of natural ecosystem services and the economic values, in terms real-world currencies and capital valuations, of those services. One of the goals of resource economics is to assist policy makers in performing the cost-benefit analy-

sis of various plans of action or inaction with regard to the natural world. The value of an ecosystem service is determined by calculating what it would cost to perform the same service artificially if the naturally-occurring service were disrupted or destroyed.

■ resource partitioning
process of dividing up resources in a ecosystem so that species with similar requirements (overlapping ecological niches) use the same scarce resources at different times, in different ways, or in different places.

■ resource recovery
the process of obtaining matter or energy from materials formerly discarded.

■ respiration
an ongoing process by which plants and animals take in oxygen and give out carbon dioxide.

■ restoration
the return of an ecosystem or habitat to its original community structure, natural complement of species, and natural functions.

■ restricted entry interval
the time after a pesticide application during which entry into the treated area is restricted.

■ restricted use
a pesticide may be classified for restricted use if it requires special handling because of its toxicity, and, if so, it may be applied only by trained, certified applicators or those under their direct supervision.

■ restriction enzymes
enzymes that recognize specific regions of a long DNA molecule and cut it at those points.

■ retrofit
addition of a pollution control device on an existing facility without making major changes to the generating plant. Also called backfit.

■ reuse
using a product or component of municipal solid waste in its original form more than once; e.g., refilling a glass bottle that has been returned or using a coffee can to hold nuts and bolts.

■ **reverse osmosis**

a treatment process used in water systems by adding pressure to force water through a semi-permeable membrane. Reverse osmosis removes most drinking water contaminants. Also used in wastewater treatment. Large-scale reverse osmosis plants are being developed.

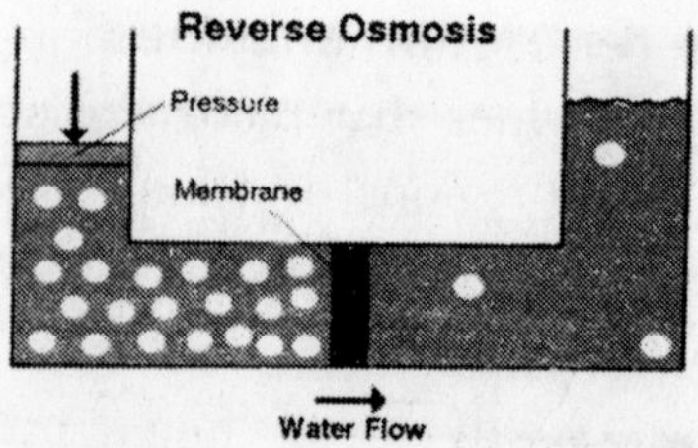

■ **reversible effect**

an effect which is not permanent; especially adverse effects which diminish when exposure to a toxic chemical stops.

■ **ribonucleic acid (RNA)**

a molecule that carries the genetic message from DNA to a cellular protein-producing mechanism.

■ **rill**

a small channel eroded into the soil by surface runoff; can be easily smoothed out or obliterated by normal tillage.

■ **ringlemann chart**

a series of shaded illustrations used to measure the opacity of air pollution emissions, ranging from light grey through black; used to set and enforce emissions standards.

■ **riparian**

having to do with the edges of streams or rivers.

■ **riparian habitat**

areas adjacent to rivers and streams with a differing density, diversity, and productivity of plant and animal species relative to nearby uplands.

■ **riparian rights**

entitlement of a land owner to certain uses of water on or bordering the property, including the right to prevent diversion or misuse of upstream waters. Generally a matter of state law.

■ **risk**

a measure of the probability that damage to life, health, property, and/or the environment will occur as a result of a given hazard.

■ **risk (adverse) for en-**

dangered species
risk to aquatic species if anticipated pesticide residue levels equal one-fifth of LD10 or one-tenth of LC50; risk to terrestrial species if anticipated pesticide residue levels equal one-fifth of LC10 or one-tenth of LC50.

■ **risk assessment**
qualitative and quantitative evaluation of the risk posed to human health and/or the environment by the actual or potential presence and/or use of specific pollutants.

■ **risk characterisation**
the last phase of the risk assessment process that estimates the potential for adverse health or ecological effects to occur from exposure to a stressor and evaluates the uncertainty involved.

■ **risk communication**
the exchange of information about health or environmental risks among risk assessors and managers, the general public, news media, interest groups, etc.

■ **risk estimate**
a description of the probability that organisms exposed to a specific dose of a chemical or other pollutant will develop an adverse response, e.g., cancer.

■ **risk factor**
characteristics (e.g., race, sex, age, obesity) or variables (e.g., smoking, occupational exposure level) associated with increased probability of a toxic effect.

■ **risk for non-endangered species**
risk to species if anticipated pesticide residue levels are equal to or greater than LC50.

■ **risk management**
the process of evaluating and selecting alternative regulatory and non-regulatory responses to risk. The selection process necessarily requires the consideration of legal, economic, and behavioural factors.

■ **risk-based targeting**
the direction of resources to those areas that have been identified as having the highest potential or actual adverse

effect on human health and/ or the environment.

risk-specific dose
the dose associated with a specified risk level.

river basin
the land area drained by a river and its tributaries.

rodenticide
a chemical or agent used to destroy rats or other rodent pests, or to prevent them from damaging food, crops, etc.

rotary kiln incinerator
an incinerator with a rotating combustion chamber that keeps waste moving, thereby allowing it to vaporise for easier burning.

rough fish
fish not prised for sport or eating, such as gar and suckers. Most are more tolerant of changing environmental conditions than are game or food species.

Sucker Family -- White Sucker

route of exposure
the avenue by which a chemical comes into contact with an organism, e.g., inhalation, ingestion, dermal contact, injection.

rubbish
solid waste, excluding food waste and ashes, from homes, institutions, and workplaces.

running losses
evaporation of motor vehicle fuel from the fuel tank while the vehicle is in use.

runoff
surface water effluent (usually from precipitation but may be from human activities such as irrigation) that moves too quickly to be absorbed into the ground. It flows down contour gradients to enter stream and river systems, carrying with it anything light enough to be borne in the volume of water, which may be light after a small rain or tremendous in the wake of a storm, when even large boulders and trees get swept up in the runoff. When runoff travels over deforested or unplanted agricultural lands, it carries away

large quantities of topsoil . Runoff from agricultural areas often carries heavy doses of biocides , fertilise rs, and other nutrients, which can lead to eutrophication when introduced into aquatic systems.

■ sacrifical anode
an easily corroded material deliberately installed in a pipe or intake to give it up (sacrifice it) to corrosion while the rest of the water supply facility remains relatively corrosion-free.

■ safe
condition of exposure under which there is a practical certainty that no harm will result to exposed individuals.

■ safe water
water that does not contain harmful bacteria, toxic materials, or chemicals, and is con sidered safe for drinking even if it may have taste, odour, colour, and certain mineral problems.

■ safe yield
the annual amount of water that can be taken from a source of supply over a period of years without depleting that source beyond its ability to be replenished naturally in "wet years."

■ safener
a chemical added to a pesticide to keep it from injuring plants.

■ salinisation
accumulation of salts in soil that can eventually make the soil unable to support plant growth.

■ salinity
a measure of the salt concentration of water. Higher salinity means more dissolved salts.

■ salt water intrusion
the invasion of fresh surface or ground water by salt water. If it comes from the ocean it may be called sea water intrusion.

■ salts
minerals that water picks up as it passes through the air, over and under the ground, or from households and industry.

■ salvage
the utilisation of waste materials.

■ **sampling frequency**
the interval between the collection of successive samples.

■ **sand filters**
devices that remove some suspended solids from sewage. Air and bacteria decompose additional wastes filtering through the sand so that cleaner water drains from the bed.

■ **sanitary landfill**
see **landfills.**

■ **sanitary sewers**
underground pipes that carry off only domestic or industrial waste, not storm water.

■ **sanitary survey**
an on-site review of the water sources, facilities, equipment, operation and maintenance of a public water system to evaluate the adequacy of those elements for producing and distributing safe drinking water.

■ **sanitary water (also known as gray water)**
water discharged from sinks, showers, kitchens, or other non-industrial operations, but not from commodes.

■ **sanitation**
control of physical factors in the human environment that could harm development, health, or survival.

■ **saprolite**
a soft, clay-rich, thoroughly decomposed rock formed in place by chemical weathering of igneous or metamorphic rock. Forms in humid, tropical, or subtropical climates.

■ **saprophyte**
organism which feeds on dead and decaying organisms, allowing the nutrients to be recycled into the ecosystem. Fungi and bacteria are two groups with many important saprophytes.

■ **saturated zone**
the area below the water table where all open spaces are

filled with water under pressure equal to or greater than that of the atmosphere.

■ **saturation**

the condition of a liquid when it has taken into solution the maximum possible quantity of a given substance at a given temperature and pressure.

■ **savannah**

a land which is without trees but with much grass either tall or short (such as the African savannah).

■ **scrap**

materials discarded from manufacturing operations that may be suitable for reprocessing.

■ **scrap metal processor**

intermediate operating facility where recovered metal is sorted, cleaned of contaminants, and prepared for recycling.

■ **screening**

use of screens to remove coarse floating and suspended solids from sewage.

■ **screening risk assessment**

a risk assessment performed with few data and many assumptions to identify exposures that should be evaluated more carefully for potential risk.

■ **scrub**

plants, such as small trees and shrubs, that usually have many stems, unlike trees which have one main trunk.

■ **scrubber**

an air pollution device that uses a spray of water or reactant or a dry process to trap pollutants in emissions.

■ **sea level**

the level of the surface of the ocean.

■ **saturation**

any large photosynthetic protist, including rhodophytes and kelps. Seaweeds are not true plants, but like plants they can make their own food. More info?

■ second law of thermodynamics
in any conversion of heat energy to useful work, some of the initial energy input is always degraded to lower quality, more dispersed, less useful energy usually low-temperature heat that flows into the environment; every energy system has "leaks" and looses energy or heat to attenuation.

■ secondary effect
action of a stressor on supporting components of the ecosystem, which in turn impact the ecological component of concern.

■ secondary materials
materials that have been manufactured and used at least once and are to be used again.

■ secondary standards
national ambient air quality standards designed to protect welfare, including effects on soils, water, crops, vegetation, man-made (anthropogenic) materials, animals, wildlife, weather, visibility, and climate; damage to property; transportation hazards; economic values, and personal comfort and well-being.

■ secondary treatment
the second step in most publicly owned waste treatment systems in which bacteria consume the organic parts of the waste. It is accomplished by bringing together waste, bacteria, and oxygen in trickling filters or in the activated sludge process. This treatment removes floating and settleable solids and about 90 percent of the oxygen-demanding substances and suspended solids. Disinfection is the final stage of secondary treatment. (See primary, tertiary treatment.).

■ second-order consumers
animals that eat first-order consumers.

■ secure chemical landfill
See landfills.

■ secure maximum contaminant level
maximum permissible level of a contaminant in water delivered to the free flowing outlet of the ultimate user, or of contamination resulting

from corrosion of piping and plumbing caused by water quality.

■ **sediment yield**
the quantity of sediment arriving at a specific location.

■ **sedimentation**
letting solids settle out of wastewater by gravity during treatment.

■ **sedimentation tanks**
wastewater tanks in which floating wastes are skimmed off and settled solids are removed for disposal.

■ **sediments**
soil, sand, and minerals washed from land into water, usually after rain. They pile up in reservoirs, rivers and harbors, destroying fish and wildlife habitat, and clouding the water so that sunlight cannot reach aquatic plants. Careless farming, mining, and building activities will expose sediment materials, allowing them to wash off the land after rainfall.

■ **seed protectant**
a chemical applied before planting to protect seeds and seedlings from disease or insects.

■ **seedbank**
a facility designed for the *ex situ* conservation of individual plan varieties through seed preservation and storage.

■ **seepage**
percolation of water through the soil from unlined canals, ditches, laterals, watercourses, or water storage facilities.

■ **selection**
natural selection is the differential contribution of offspring to the next generation by various genetic types belonging to the same populations. Artificial selection is the intentional manipulation by man of the fitness of individuals in a population to produce a desired evolutionary response.

■ **selective pesticide**
a chemical designed to affect only certain types of pests, leaving other plants and animals unharmed.

■ **semiarid**
having a climate that is dry, but not as dry as a desert.

■ semi-confined aquifer
an aquifer partially confined by soil layers of low permeability through which recharge and discharge can still occur.

■ semivolatile organic compounds
organic compounds that volatilise slowly at standard temperature (20 degrees C and 1 atm pressure).

■ senescence
the aging process. Sometimes used to describe lakes or other bodies of water in advanced stages of eutrophication. Also used to describe plants and animals.

■ septic system
an on-site system designed to treat and dispose of domestic sewage. A typical septic system consists of tank that receives waste from a residence or business and a system of tile lines or a pit for disposal of the liquid effluent (sludge) that remains after decomposition of the solids by bacteria in the tank and must be pumped out periodically.

■ septic tank
an underground storage tank for wastes from homes not connected to a sewer line. Waste goes directly from the home to the tank. (See septic system.).

■ service connector
the pipe that carries tap water from a public water main to a building.

■ service line sample
a one-litre sample of water that has been standing for at least 6 hours in a service pipeline and is collected according to federal regulations.

■ service pipe
the pipeline extending from the water main to the building served or to the consumer's system.

■ set-back
setting a thermometer to a lower temperature when the building is unoccupied to reduce consumption of heating energy. Also refers to setting the thermometer to a higher temperature during unoccupied periods in the cooling season.

■ **settleable solids**
material heavy enough to sink to the bottom of a wastewater treatment tank.

■ **settling chamber**
a series of screens placed in the way of flue gases to slow the stream of air, thus helping gravity to pull particles into a collection device.

■ **settling tank**
a holding area for wastewater, where heavier particles sink to the bottom for removal and disposal.

■ **sewage**
the waste and wastewater produced by residential and commercial sources and discharged into sewers.

■ **sewage lagoon**
see **lagoon.**

■ **sewage sludge**
sludge produced at a Publicly Owned Treatment Works, the disposal of which is regulated under the Clean Water Act.

■ **sewer**
a channel or conduit that carries wastewater and stormwater runoff from the source to a treatment plant or receiving stream. "Sanitary" sewers carry household, industrial, and commercial waste. "Storm" sewers carry runoff from rain or snow. "Combined" sewers handle both.

■ **sewerage**
the entire system of sewage collection, treatment, and disposal.

■ **shading coefficient**
the amount of the sun's heat transmitted through a given window compared with that of a standard 1/8- inch-thick single pane of glass under the same conditions.

■ **sharps**
hypodermic needles, syringes (with or without the attached needle), Pasteur pipettes, scalpel blades, blood vials, needles with attached tubing, and culture dishes used in animal or human patient care or treatment, or in medical, re search or industrial laboratories. Also included are other types of broken or unbroken glassware that were in contact with infectious agents, such as used slides and cover slips, and unused hypodermic and

suture needles, syringes, and scalpel blades.

shock load

the arrival at a water treatment plant of raw water containing unusual amounts of algae, colloidal matter. colour, suspended solids, turbidity, or other pollutants.

short-circuiting

when some of the water in tanks or basins flows faster than the rest; may result in shorter contact, reaction, or settling times than calculated or presumed.

sick building syndrome

building whose occupants experience acute health and/or comfort effects that appear to be linked to time spent therein, but where no specific illness or cause can be identified. Complaints may be localised in a particular room or zone, or may spread throughout the building.

signal

the volume or product-level change produced by a leak in a tank.

signal words

the words used on a pesticide label—Danger, Warning, Caution—to indicate level of toxicity.

significant deterioration

pollution resulting from a new source in previously "clean" areas. (See prevention of significant deterioration.).

significant municipal facilities

those publicly owned sewage treatment plants that discharge a million gallons per day or more and are therefore considered by states to have the potential to substantially affect the quality of receiving waters.

significant non-compliance

see **significant violations.**

significant potential source of contamination

a facility or activity that stores, uses, or produces compounds with potential for significant contaminating impact if released into the source water of a public water supply.

■ **significant violations**
violations by point source dischargers of sufficient magnitude or duration to be a regulatory priority.

■ **silt**
sedimentary materials composed of fine or intermediate-sised mineral particles.

■ **silviculture**
management of forest land for timber.

■ **single-breath canister**
small one-liter canister designed to capture a single breath. Used in air pollutant ingestion research.

■ **sink**
place in the environment where a compound or material collects.

■ **sinking**
controlling oil spills by using an agent to trap the oil and sink it to the bottom of the body of water where the agent and the oil are biodegraded.

■ **site assessment program**
a means of evaluating hazardous waste sites through preliminary assessments and site inspections to develop a Hazard Ranking System score.

■ **site safety plan**
a crucial element in all removal actions, it includes information on equipment being used, precautions to be taken, and steps to take in the event of an on-site emergency.

■ **siting**
the process of choosing a location for a facility.

■ **skimming**
using a machine to remove oil or scum from the surface of the water.

■ **slow sand filtration**
passage of raw water through a bed of sand at low velocity, resulting in substantial removal of chemical and biological contaminants.

■ **sludge**
a semi-solid residue from any of a number of air or water treatment processes; can be a hazardous waste.

■ **sludge digester**
tank in which complex organic substances like sewage sludges are biologically

dredged. During these reactions, energy is released and much of the sewage is converted to methane, carbon dioxide, and water.

■ **slurry**
a watery mixture of insoluble matter resulting from some pollution control techniques.

■ **smelter**
a facility that melts or fuses ore, often with an accompanying chemical change, to separate its metal content. Emissions cause pollution. "Smelting" is the process involved.

■ **smog**
air pollution typically associated with oxidants. (See photochemical smog.).

■ **smoke**
particles suspended in air after incomplete combustion.

■ **social group**
a small population that lives and travels together and in some ways depends on each other for its well-being.

■ **soft detergents**
cleaning agents that break down in nature.

■ **soft water**
any water that does not contain a significant amount of dissolved minerals such as salts of calcium or magnesium.

■ **soil adsorption field**
a sub-surface area containing a trench or bed with clean stones and a system of piping through which treated sewage may seep into the surrounding soil for further treatment and disposal.

■ **soil and water conservation practices**
control measures consisting of managerial, vegetative, and structural practices to reduce the loss of soil and water.

■ **soil conditioner**
an organic material like humus or compost that helps soil absorb water, build a bacterial community, and take up mineral nutrients.

■ **soil erodibility**
an indicator of a soil's susceptibility to raindrop impact, runoff, and other erosive processes.

■ soil erosion
the loss of topsoil through silt-laden runoff , strong winds, or other forces that transport soil away from its natural location.

■ soil gas
gaseous elements and compounds in the small spaces between particles of the earth and soil. Such gases can be moved or driven out under pressure.

■ soil moisture
the water contained in the pore space of the unsaturated zone.

■ soil sterilant
a chemical that temporarily or permanently prevents the growth of all plants and animals,

■ solar energy
energy from the sun.

■ solder
metallic compound used to seal joints between pipes. Until recently, most solder contained 50 percent lead. Use of solder containing more than 0.2 percent lead in pipes carrying drinking water is now prohibited.

■ sole-source aquifer
an aquifer that supplies 50-percent or more of the drinking water of an area.

■ solid waste
non-liquid, non-soluble materials ranging from municipal garbage to industrial wastes that contain complex and sometimes hazardous substances. Solid wastes also include sewage sludge, agricultural refuse, demolition wastes, and mining residues. Technically, solid waste also refers to liquids and gases in containers.

■ solid waste disposal
the final placement of refuse that is not salvaged or recycled.

■ solid waste management
supervised handling of waste materials from their source through recovery processes to disposal.

■ solidification and

stabilisation
removal of wastewater from a waste or changing it chemically to make it less permeable and susceptible to transport by water.

■ **solubility**
the amount of mass of a compound that will dissolve in a unit volume of solution. Aqueous Solubility is the maximum concentration of a chemical that will dissolve in pure water at a reference temperature.

■ **solution**
a mixture made by dissolving a substance in a liquid, such as water.

■ **soot**
carbon dust formed by incomplete combustion.

■ **sorption**
the action of soaking up or attracting substances; process used in many pollution control systems.

■ **source area**
the location of liquid hydrocarbons or the zone of highest soil or groundwater concentrations, or both, of the chemical of concern.

■ **source characterisation measurements**
measurements made to estimate the rate of release of pollutants into the environment from a source such as an incinerator, landfill, etc.

■ **source reduction**
reducing the amount of materials entering the waste stream from a specific source by redesigning products or patterns of production or consumption (e.g., using returnable beverage containers). Synonymous with waste reduction.

■ **source separation**
segregating various wastes at the point of generation (e.g., separation of paper, metal and glass from other wastes to make recycling simpler and more efficient).

■ **source-water protection area**
the area delineated by a state for a Public Water Supply or including numerous such suppliers, whether the source is

ground water or surface water or both.

■ **southern hemisphere**
the area of the earth below the equator.

■ **southern temperate zone**
the region between latitudes 23.5°S and 66.5°S.

■ **sparge or sparging**
injection of air below the water table to strip dissolved volatile organic compounds and/or oxygenate ground water to facilitate aerobic biodegradation of organic compounds.

■ **special waste**
items such as household hazardous waste, bulky wastes (refrigerators, pieces of furniture, etc.) tires, and used oil.

■ **specialist**
organism which has adopted a lifestyle specific to a particular set of conditions. Contrast with generalist.

■ **specialist species**
species with a narrow ecological niche. They may be able to live in only one type of habitat, tolerate only a narrow range of climatic or other environmental conditions, or they may use only one or a few types of food.

■ **speciation**
formation of two species from one species as a result of divergent natural selection in response to changes in environmental conditions; usually takes thousands or tens of thousands of years.

■ **species**
1. a reproductively isolated aggregate of interbreeding organisms having common attributes and usually designated by a common name.2. An organism belonging to belonging to such a category.

■ **species diversity**
a function of the distribution and abundance of species. Approximately synonymous with Species Richness. In more technical literature, includes considerations of the evenness of species abundances. An ecosystem is said to be more diverse, according to the more technical definition, if species present have equal population sizes and

less diverse if many species are rare and some are very common.

■ species richness
the number of species within a region. A term commonly used as a measure of Species Diversity, but technically only one aspect of diversity.

■ specific conductance
rapid method of estimating the dissolved solid content of a water supply by testing its capacity to carry an electrical current.

■ specific yield
the amount of water a unit volume of saturated permeable rock will yield when drained by gravity.

■ spoil
dirt or rock removed from its original location—destroying the composition of the soil in the process—as in strip-mining, dredging, or construction.

■ sprawl
unplanned development of open land.

■ spray tower scrubber
a device that sprays alkaline water into a chamber where acid gases are present to aid in Neutralising the gas.

■ spring
ground water seeping out of the earth where the water table intersects the ground surface.

■ spring melt/thaw
the process whereby warm temperatures melt winter snow and ice. Because various forms of acid deposition may have been stored in the frozen water, the melt can result in abnormally large amounts of acidity entering streams and rivers, sometimes causing fish kills.

■ stabilisation
conversion of the active organic matter in sludge into inert, harmless material.

■ stabilisation ponds
see **lagoon.**

■ stability
a function of several characteristics of community or ecosystem dynamics, including the degree of population fluctuations, the community's resistance to disturbances, the speed of recovery from dis-

turbances, and the persistence of the community's composition through time.

stable air
a motionless mass of air that holds, instead of dispersing, pollutants.

stack
a chimney, smokestack, or vertical pipe that discharges used air.

stack effect
flow of air resulting from warm air rising, creating a positive pressure area at the top of a building and negative pressure area at the bottom. This effect can overpower the mechanical system and disrupt building ventilation and air circulation.

stack gas
see **flue gas.**

stage ii controls
systems placed on service station gasoline pumps to control and capture gasoline vapours during refuelling.

stagnation
lack of motion in a mass of air or water that holds pollutants in place.

stakeholder
any organisation, governmental entity, or individual that has a stake in or may be impacted by a given approach to environmental regulation, pollution prevention, energy conservation, etc.

standard sample
the part of finished drinking water that is examined for the presence of coliform bacteria.

state environmental goals and indication project
program to assist state environmental agencies by providing technical and financial assistance in the development of environmental goals and indicators.

static water depth
the vertical distance from the centreline of the pump discharge down to the surface level of the free pool while no water is being drawn from the pool or water table.

static water level
1. elevation or level of the water table in a well when the pump is not operating. 2. The level or elevation to which

water would rise in a tube connected to an artesian aquifer or basin in a conduit under pressure.

■ stationary source
a fixed-site producer of pollution, mainly power plants and other facilities using industrial combustion processes.

■ sterilisation
the removal or destruction of all microorganisms, including pathogenic and other bacteria, vegetative forms, and spores.

■ stomata
tiny pores on the surface of plant leaves that can open and close to take in and give out water vapour.

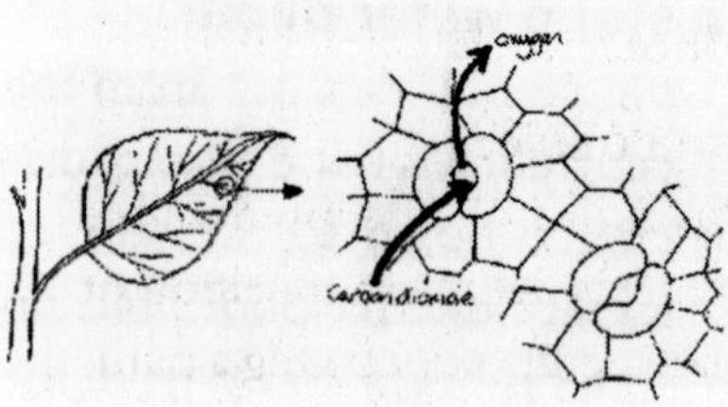

■ storage
temporary holding of waste pending treatment or disposal, as in containers, tanks, waste piles, and surface impoundments.

■ storm sewer
a system of pipes (separate from sanitary sewers) that carries water runoff from buildings and land surfaces.

■ stratification
separating into layers.

■ stratigraphy
study of the formation, composition, and sequence of sediments, whether consolidated or not.

■ stratonimbus
stratus clouds which produce a steady rainfall.

■ stratosphere
the portion of the atmosphere 10-to-25 miles above the earth's surface.

■ stratus
clouds which are produced by stable air and looks like an even blanket.

■ stressors
physical, chemical, or biological entities that can induce adverse effects on ecosystems or human health.

■ strip-cropping
growing crops in a systematic arrangement of strips or

bands that serve as barriers to wind and water erosion.

strip-mining
a process that uses machines to scrape soil or rock away from mineral deposits just under the earth's surface.

structural deformation
distortion in walls of a tank after liquid has been added or removed.

subcanopy layer
a forest's third layer, formed by the leaves and branches of shorter trees under the canopy layer.

subchronic
of intermediate duration, usually used to describe studies or periods of exposure lasting between 5 and 90 days.

subchronic exposure
multiple or continuous exposures lasting for approximately ten percent of an experimental species lifetime, usually over a three-month period.

submerged aquatic vegetation
vegetation that lives at or below the water surface; an important habitat for young fish and other aquatic organisms.

subspecies
a subdivision of a species; a population or series of populations occupying a discrete range of differing genetically from other subspecies of the same species.

substrate
"Supporting surface" on which an organism grows. The substrate may simply provide structural support, or may provide water and nutrients. A substrate may be inorganic, such as rock or soil, or it may be organic, such as wood.

subwatershed
topographic perimeter of the catchment area of a stream tributary.

succession
the more or less predictable changes in the composition of communities following a natural or human disturbance. For example, after a gap is made in a forest by logging, clearing, fire, or treefall, the initial (or "pioneer") species are often fast-growing and shade-intolerant. These species are

eventually replaced by shade-tolerant species that can grown beneath the pioneers. If a community is not further disturbed, the outcome of the successional sequence may be a so-called *climax* community whose composition is unchanging. In practice, many communities are frequently disturbed and may never reach a climax composition.

■ **succulent plants**
plants that have thick, fleshy leaves or stems for storing water.

■ **sulphur cycle**
cyclic movement of sulphur in different chemical forms, from the environment to organisms and then back to the environment.

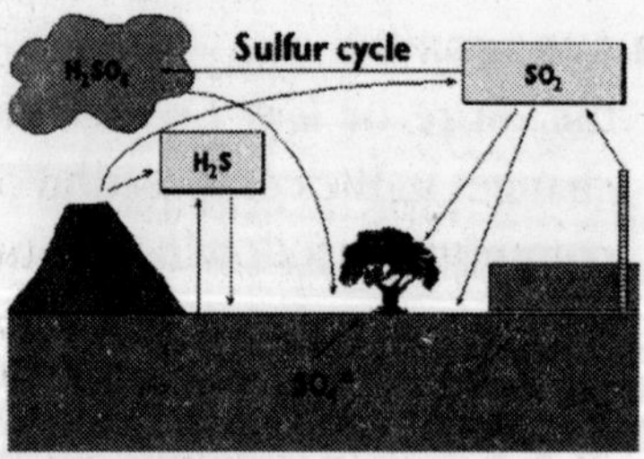

■ **sulphur dioxide (SO_2)**
a pungent, colourless, gas formed primarily by the combustion of fossil fuels; becomes a pollutant when present in large amounts.

■ **sump**
a pit or tank that catches liquid runoff for drainage or disposal.

■ **sunlight zone**
the upper 488 feet (150 m) of the ocean, where sunlight penetrates and where about 90 percent of all marine life live.

■ **superchlorination**
chlorination with doses that are deliberately selected to produce water free of combined residuals so large as to require dechlorination.

■ **supercritical water**
a type of thermal treatment using moderate temperatures and high pressures to enhance the ability of water to break down large organic molecules into smaller, less toxic ones. Oxygen injected during this process combines with simple organic compounds to form carbon dioxide and water.

■ **superorganism**
an organism, such as coral or the Portuguese man-of-war, that appears to be one organism, but in fact is a number

of colonial animals joined together.

■ **supplemental registration**
an arrangement whereby a registrant licenses another company to market its pesticide product under the second company's registration.

■ **supplier of water**
any person who owns or operates a public water supply.

■ **surface impoundment**
treatment, storage, or disposal of liquid hazardous wastes in ponds.

■ **surface runoff**
precipitation, snow melt, or irrigation water in excess of what can infiltrate the soil surface and be stored in small surface depressions; a major transporter of non-point source pollutants in rivers, streams, and lakes..

■ **surface uranium mines**
strip mining operations for removal of uranium bearing ore.

■ **surface water**
all water naturally open to the atmosphere (rivers, lakes, reservoirs, ponds, streams, impoundments, seas, estuaries, etc.).

■ **surface-water treatment rule**
rule that specifies maximum contaminant level goals for Giardia lamblia, viruses, and Legionella and promulgates filtration and disinfection requirements for public water systems using surface-water or ground-water sources under the direct influence of surface water. The regulations also specify water quality, treatment, and watershed protection criteria under which filtration may be avoided.

■ **surfacing ACM**
asbestos-containing material that is sprayed or trawled on or otherwise applied to surfaces, such as acoustical plaster on ceilings and fireproofing materials on structural members.

■ **surfacing material**
material sprayed or trawled onto structural members (beams, columns, or decking) for fire protection; or on ceilings or walls for fireproofing, acoustical or decorative purposes. Includes textured plas-

ter, and other textured wall and ceiling surfaces.

■ surfactant
a detergent compound that promotes lathering.

■ surrogate data
data from studies of test organisms or a test substance that are used to estimate the characteristics or effects on another organism or substance.

■ surveillance system
a series of monitoring devices designed to check on environmental conditions.

■ susceptibility analysis
an analysis to determine whether a Public Water Supply is subject to significant pollution from known potential sources.

■ suspect material
building material suspected of containing asbestos; e.g., surfacing material, floor tile, ceiling tile, thermal system insulation.

■ suspended loads
specific sediment particles maintained in the water column by turbulence and carried with the flow of water.

■ suspended solids
small particles of solid pollutants that float on the surface of, or are suspended in, sewage or other liquids. They resist removal by conventional means.

■ suspension culture
cells growing in a liquid nutrient medium.

■ sustainability
ability of a system to survive for some specified (finite) time.

■ sustainable agriculture
method of growing crops and raising livestock based on organic fertilise rs, soil conservation, water conservation, biological control of pests, and minimal use of non-renewable fossil-fuel energy.

■ sustainable development
development that meets the needs and aspirations of the current generation without compromising the ability to meet those of future generations.

■ sustainable living
taking no more potentially renewable resources from the natural world than can be replenished naturally and not

overloading the capacity of the environment to cleanse and renew itself by natural processes.

■ sustainable society

a society that manages its economy and population size without doing irreparable environmental harm by overloading the planet's ability to absorb environmental wastes, replenish its resources, and sustain human and other forms of life over a specified period-usually hundreds to thousands of years. During this period it satisfies the needs of its people without depleting earth capital and thereby jeopardising the prospects of current and future generations of humans and other species.

■ sustainable system

a system that survives and functions over some specified (finite) time; a system that attains its full expected lifetime.

■ sustainable yield

highest rate at which a potentially renewable resource can be used without reducing its available supply throughout the world or in a particular area.

■ swamp

a type of wetland dominated by woody vegetation but without appreciable peat deposits. Swamps may be fresh or salt water and tidal or non-tidal. (See wetlands.).

■ symbiont

see symbiosis .

■ symbiosis

literally means "living together" in Latin. Any intimate relationship or association between members of two or more species. The members of the relationship are symbionts. Obligate symbionts rely so heavily on the relationship that they cannot feed, reproduce, or perform some other crucial life function in the absence of their symbiotic partner(s). There are three main categories of symbiosis commensalism , mutualism , and parasitism , with some degree of blending at the edges of these definitions in many cases.

■ symbiotic

refers to a component or member of a system of symbiosis. "These organisms have a symbiotic relationship."

■ synergism
an interaction of two or more chemicals that results in an effect greater than the sum of their separate effects.

■ Synthetic Organic Chemicals (SOCs)
man-made (anthropogenic) organic chemicals. Some SOCs are volatile; others tend to stay dissolved in water instead of evapourating.

■ system with a single service connection
a system that supplies drinking water to consumers via a single service line.

■ systematics
the study of the historical evolutionary and genetic relationships among organisms and of their phenotypic similarities and differences.

■ systemic pesticide
a chemical absorbed by an organism that interacts with the organism and makes the organism toxic to pests.

■ tail water
the runoff of irrigation water from the lower end of an irrigated field.

■ tailings
residue of raw material or waste separated out during the processing of crops or mineral ores.

■ tailpipe standards
emissions limitations applicable to mobile source engine exhausts.

■ tampering
adjusting, negating, or removing pollution control equipment on a motor vehicle.

■ taxon (pl. taxa)
the named classification unit (e.g. *Homo sapiens*, Hominidae, or Mammalia) to which individuals, or sets of species, are assigned. *Higher taxa* are those above the species level.

■ taxonomy
the naming and assignment of organisms to taxa.

■ technical-grade active ingredient (tga)
a pesticide chemical in pure form as it is manufactured prior to being formulated into an end-use product (e.g. wet-

table powders, granules, emulsifiable concentrates). Registered manufactured products composed of such chemicals are known as Technical Grade Products.

■ **technology-based limitations**
industry-specific effluent limitations based on best available preventive technology applied to a discharge when it will not cause a violation of water quality standards at low stream flows. Usually applied to discharges into large rivers.

■ **technology-based standards**
industry-specific effluent limitations applicable to direct and indirect sources which are developed on a category-by-category basis using statutory factors, not including water-quality effects.

■ **temperate**
region in which the climate undergoes seasonal change in temperature and moisture. Temperate regions of the earth lie primarily between 30 and 60 degrees latitude in both hemispheres.

■ **teratogenesis**
the introduction of nonhereditary birth defects in a developing foetus by exogenous factors such as physical or chemical agents acting in the womb to interfere with normal embryonic development.

■ **terracing**
dikes built along the contour of sloping farm land that hold runoff and sediment to reduce erosion.

■ **terrestrial**
living on land, as opposed to marine or aquatic.

■ **tertiary treatment**
advanced cleaning of wastewater that goes beyond the secondary or biological stage, removing nutrients such as phosphorus, nitrogen, and most BOD and suspended solids.

■ **theoretical maximum residue contribution**
the theoretical maximum amount of a pesticide in the daily diet of an average person. It assumes that the diet is composed of all food items for which there are tolerance-level residues of the pesticide. The TMRC is expressed

as milligrams of pesticide/kilograms of body weight/day.

therapeutic index
the ratio of the dose required to produce toxic or lethal effects to the dose required to produce nonadverse or therapeutic response.

thermal pollution
discharge of heated water from industrial processes that can kill or injure aquatic organisms.

thermal stratification
the formation of layers of different temperatures in a lake or reservoir.

thermal system insulation (TSI)
asbestos-containing material applied to pipes, fittings, boilers, breeching, tanks, ducts, or other interior structural components to prevent heat loss or gain or water condensation.

thermal treatment
use of elevated temperatures to treat hazardous wastes.

thermocline
the middle layer of a thermally stratified lake or reservoir. In this layer, there is a rapid decrease in temperatures in a lake or reservoir.

third order consumers
animals that eat first- and/or second-order consumers.

threshold
the lowest dose of a chemical at which a specified measurable effect is observed and below which it is not observed.

threshold level
time-weighted average pollutant concentration values, exposure beyond which is likely to adversely affect human health. (See environmental exposure).

threshold limit value (TLV)
the concentration of an airborne substance to which an average person can be repeatedly exposed without adverse effects. TLVs may be expressed in three ways (1) TLV-TWA—Time weighted average, based on an allowable exposure averaged over a normal 8-hour workday or 40-hour work- week; (2) TLV-STEL—Short-term exposure limit or maximum concentration for a brief specified pe-

riod of time, depending on a specific chemical (TWA must still be met); and (3) TLV-C—Ceiling Exposure Limit or maximum exposure concentration not to be exceeded under any circumstances. (TWA must still be met.).

■ **threshold odour**
see **Odour threshold.**

■ **thropic levels**
a functional classification of species that is based on feeding relationships (e.g. generally aquatic and terrestrial green plants comprise the first thropic level, and herbivores comprise the second.).

■ **tidal marsh**
low, flat marshlands traversed by channels and tidal hollows, subject to tidal inundation; normally, the only vegetation present is salt-tolerant bushes and grasses. (See wetlands.).

■ **tillage**
ploughing, seedbed preparation, and cultivation practices.

■ **time-weighted average (TWA)**
in air sampling, the average air concentration of contaminants during a given period.

■ **tire processor**
intermediate operating facility where recovered tires are processed in preparation for recycling.

■ **tires**
as used in recycling, passenger car and truck tires (excludes airplane, bus, motorcycle and special service military, agricultural, off-the-road and-slow speed industrial tires). Car and truck tires are recycled into rubber products such as trash cans, storage containers, rubberised asphalt or used whole for playground and reef construction.

■ **tolerance petition**
a formal request to establish a new tolerance or modify an existing one.

■ **tonnage**
the amount of waste that a landfill accepts, usually expressed in tons per month. The rate at which a landfill accepts waste is limited by the landfill's permit.

■ **top consumer**
an organism at the top of a food chain.

■ topography
the physical features of a surface area including relative elevations and the position of natural and man-made (anthropogenic) features.

■ total dissolved phosphorous
the total phosphorous content of all material that will pass through a filter, which is determined as orthophosphate without prior digestion or hydrolysis. Also called soluble P. or ortho P.

■ total dissolved solids (TDS)
all material that passes the standard glass river filter; now called total filtrable residue. Term is used to reflect salinity.

■ total petroleum hydrocarbons (TPH)
measure of the concentration or mass of petroleum hydrocarbon constituents present in a given amount of soil or water. The word "total" is a misnomer—few, if any, of the procedures for quantifying hydrocarbons can measure all of them in a given sample. Volatile ones are usually lost in the process and not quantified and non-petroleum hydrocarbons sometimes appear in the analysis.

■ total recovered petroleum hydrocarbon
a method for measuring petroleum hydrocarbons in samples of soil or water.

■ total suspended particles (TSP)
a method of monitoring airborne particulate matter by total weight.

■ total suspended solids (TSS)
a measure of the suspended solids in wastewater, effluent, or water bodies, determined by tests for "total suspended non-filterable solids.

■ toxaphene
chemical that causes adverse health effects in domestic water supplies and is toxic to fresh water and marine aquatic life.

■ toxic chemical release form
information form required of facilities that manufacture, process, or use (in quantities above a specific amount)

chemicals listed under SARA Title III.

■ toxic chemical use substitution
replacing toxic chemicals with less harmful chemicals in industrial processes.

■ toxic cloud
airborne plume of gases, vapours, fumes, or aerosols containing toxic materials.

■ toxic concentration
the concentration at which a substance produces a toxic effect.

■ toxic dose
the dose level at which a substance produces a toxic effect.

■ toxic pollutants
materials that cause death, disease, or birth defects in organisms that ingest or absorb them. The quantities and exposures necessary to cause these effects can vary widely.

■ toxic substance
a chemical or mixture that may present an unreasonable risk of injury to health or the environment.

■ toxic waste
a waste that can produce injury if inhaled, swallowed, or absorbed through the skin.

■ toxicant
a harmful substance or agent that may injure an exposed organism.

■ toxicity
the degree to which a substance or mixture of substances can harm humans or animals. *Acute toxicity* involves harmful effects in an organism through a single or short-term exposure. *Chronic toxicity* is the ability of a substance or mixture of substances to cause harmful effects over an extended period, usually upon repeated or continuous exposure sometimes lasting for the entire life of the exposed organism. *Subchronic toxicity* is the ability of the substance to cause effects for more than one year but less than the life time of the exposed organism.

■ toxicity assessment
characterisation of the toxicological properties and effects of a chemical, with special emphasis on establishment of dose-response characteristics.

toxicity testing
biological testing (usually with an invertebrate, fish, or small mammal) to determine the adverse effects of a compound or effluent.

toxicological profile
an examination, summary, and interpretation of a hazardous substance to determine levels of exposure and associated health effects.

transboundary pollutants
air pollution that travels from one jurisdiction to another, often crossing state or international boundaries. Also applies to water pollution.

transfer station
facility where solid waste is transferred from collection vehicles to larger trucks or rail cars for longer distance transport.

transient water system
a non-community water system that does not serve 25 of the same non-residents per day for more than six months per year.

transmission lines
pipelines that transport raw water from its source to a water treatment plant, then to the distribution grid system.

transmissivity
the ability of an aquifer to transmit water.

transpiration
the loss of water into the atmosphere through the stomata of plants.

transportation control measures
steps taken by a locality to reduce vehicular emission and improve air quality by reducing or changing the flow of traffic; e.g. bus and HOV lanes, carpooling and other forms of ride-sharing, public transit, bicycle lanes.

trash
material considered worthless or offensive that is thrown away. Generally defined as dry waste material, but in common usage it is a synonym for garbage, rubbish, or refuse.

trash-to-energy plan
burning trash to produce energy.

treatability studies
tests of potential cleanup technologies conducted in a

laboratory (See bench-scale tests.).

treated regulated medical waste
medical waste treated to substantially reduce or eliminate its pathogenicity, but that has not yet been destroyed.

treated wastewater
wastewater that has been subjected to one or more physical, chemical, and biological processes to reduce its poten tial of being health hazard.

treatment
1. any method, technique, or process designed to remove solids and/or pollutants from solid waste, waste-streams, effluents, and air emissions. (2) Methods used to change the biological character or composition of any regulated medical waste so as to substantially reduce or eliminate its potential for causing disease.

treatment plant
a structure built to treat wastewater before discharging it into the environment.

tree line
the height on a mountain above which the climate is too cold for trees to grow.

tremie
device used to place concrete or grout under water.

trial burn
an incinerator test in which emissions are monitored for the presence of specific organic compounds, particulates, and hydrogen chloride.

trichloroethylene (TCE)
a stable, low boiling-point colourless liquid, toxic if inhaled. Used as a solvent or metal degreasing agent, and in other industrial applications.

trickle irrigation
method in which water drips to the soil from perforated tubes or emitters.

trickling filter
a coarse treatment system in which wastewater is trickled over a bed of stones or other material covered with bacteria that break down the organic waste and produce clean water.

■ trihalomethane (THM)
one of a family of organic compounds named as derivative of methane. THMs are generally by-products of chlorination of drinking water that contains organic material.

■ trophic level
position in the food chain, determined by the number of energy-transfer steps to that level.

■ tropical
region in which the climate undergoes little seasonal change in either temperature or rainfall. Tropical regions of the earth lie primarily between 30 degrees north and south of the equator.

■ tropical rain forest
a forest that gains more water from precipitation than it loses through evaporation. Located in the tropical zone and having an average temperature between 70¡ and 85¡F (21¡ and 29¡C) and average yearly rainfall of more than 80 inches (200 cm).

■ tropical zone
the region between latitudes 23.5°S and 23.5°N.

■ troposhpere
the layer of the atmosphere closest to the earth's surface.

■ tube settler
device using bundles of tubes to let solids in water settle to the bottom for removal by conventional sludge collection means; sometimes used in sedimentation basins and clarifiers to improve particle removal.

■ tuberculation
development or formation of small mounds of corrosion products on the inside of iron pipe. These tubercules roughen the inside of the pipe, increasing its resistance to water flow.

■ tundra
a type of treeless ecosystem dominated by lichens, mosses, grasses, and woody plants. Tundra is found at high latitudes (arctic tundra) and high altitudes (alpine tundra). Arctic tundra is underlain by permafrost and is usually water saturated. (See wetlands.).

■ turbidimeter
a device that measures the cloudiness of suspended solids

in a liquid; a measure of the quantity of suspended solids.

■ turbidity

1. haziness in air caused by the presence of particles and pollutants. 2. a cloudy condition in water due to suspended silt or organic matter.

■ twilight zone

the shadowy area of the ocean, extending from the bottom of the sunlight zone down to about 3,000 feet (1 ,000 m), where plants cannot grow and where animals are less numerous and smaller High-energy rays of sunlight.

■ ultra clean coal

coal that is washed, ground into fine particles, then chemically treated to remove sulphur, ash, silicone, and other substances; usually briquetted and coated with a sealant made from coal.

■ ultraviolet rays

radiation from the sun that can be useful or potentially harmful. UV rays from one part of the spectrum (UV-A) enhance plant life. UV rays from other parts of the spectrum (UV-B) can cause skin cancer or other tissue damage. The ozone layer in the atmosphere partly shields us from ultraviolet rays reaching the earth's surface.

■ uncertainty factor

one of several factors used in calculating the reference dose from experimental data. UFs are intended to account for (1) the variation in sensitivity among humans; (2) the uncertainty in extrapolating animal data to humans; (3) the uncertainty in extrapolating data obtained in a study that covers less than the full life of the exposed animal or human.

■ unconfined aquifer

an aquifer containing water that is not under pressure; the water level in a well is the same as the water table outside the well.

■ underground injection control

the program under the Safe Drinking Water Act that regulates the use of wells to pump fluids into the ground.

■ underground injection wells

steel- and concrete-encased shafts into which hazardous waste is deposited by force and under pressure.

underground sources of drinking water
aquifers currently being used as a source of drinking water or those capable of supplying a public water system. They have a total dissolved solids content of 10,000 milligrams per liter or less, and are not "exempted aquifers." (See exempted aquifer.).

underground storage tank
a tank located at least partially underground and designed to hold gasoline or other petroleum products or chemicals.

universe
the solar system beyond our world; see universe

unreasonable risk
under the Federal Insecticide, Fungicide, and Rodenticide Act (FIFRA), "unreasonable adverse effects" means any unreasonable risk to man or the environment, taking into account the medical, economic, social, and environmental costs and benefits of any pesticide.

unsaturated zone
the area above the water table where soil pores are not fully saturated, although some water may be present.

upper detection limit
the largest concentration that an instrument can reliably detect.

upwelling
the raising of benthic nutrients to the surface waters. This occurs in regions where the flow of water brings currents of differing temperatures together, and increases productivity of the ecosystem.

uranium mill tailings piles
former uranium ore processing sites that contain leftover radioactive materials (wastes), including radium and unrecovered uranium.

urban runoff
storm water from city streets and adjacent domestic or commercial properties that carries pollutants of various kinds into the sewer systems and receiving waters.

urea-formaldehyde foam insulation
a material once used to conserve energy by sealing crawl spaces, attics, etc.; no longer

used because emissions were found to be a health hazard.

use cluster
a set of competing chemicals, processes, and/or technologies that can substitute for one another in performing a particular function.

used oil
spent motor oil from passenger cars and trucks collected at specified locations for recycling (not included in the category of municipal solid waste).

user fee
fee collected from only those persons who use a particular service, as compared to one collected from the public in general.

utility load
the total electricity demand for a utility district.

vadose zone
the zone between land surface and the water table within which the moisture content is less than saturation (except in the capillary fringe) and pressure is less than atmospheric. Soil pore space also typically contains air or other gases. The capillary fringe is included in the vadose zone.

valued environmental attributes/components
those aspects (components / processes /functions) of ecosystems, human health, and environmental welfare considered to be important and potentially at risk from human activity or natural hazards. Similar to the term "valued environmental components" used in environmental impact assessment.

vapour capture system
any combination of hoods and ventilation system that captures or contains organic vapours so they may be directed to an abatement or recovery device.

vapour dispersion
the movement of vapour clouds in air due to wind, thermal action, gravity spreading, and mixing.

vapour plumes
flue gases visible because they contain water droplets.

vapour pressure
a measure of a substance's propensity to evaporate, vapour pressure is the force

per unit area exerted by vapour in an equilibrium state with surroundings at a given pressure. It increases exponentially with an increase in temperature. A relative measure of chemical volatility, vapour pressure is used to calculate water partition coefficients and volatilisation rate constants.

■ variance
government permission for a delay or exception in the application of a given law, ordinance, or regulation.

■ vascular plants
plants with a well-developed vascular system that transports water, minerals, sugars, and other nutrients throughout the plant body. Excludes the bryophytes: mosses, hornworts, and liverworts.

■ vector
1. an organism, often an insect or rodent, that carries disease. 2. Plasmids, viruses, or bacteria used to transport genes into a host cell. A gene is placed in the vector; the vector then "infects" the bacterium.

■ vegetation
plant life.

■ vegetative controls
non-point source pollution control practices that involve vegetative cover to reduce erosion and Minimise loss of pollutants.

■ vehicle miles travelled
a measure of the extent of motor vehicle operation; the total number of vehicle miles travelled within a specific geographic area over a given period of time.

■ ventilation rate
the rate at which indoor air enters and leaves a building. Expressed as the number of changes of outdoor air per unit of time (air changes per hour (ACH), or the rate at which a volume of outdoor air enters in cubic feet per minute (CFM).

■ ventilation/suction
the act of admitting fresh air into a space in order to replace stale or contaminated air; achieved by blowing air into the space. Similarly, suction represents the admission of fresh air into an interior space by lowering the pressure outside of the space, thereby

drawing the contaminated air outward.

■ **venturi scrubbers**
air pollution control devices that use water to remove particulate matter from emissions.

■ **vinyl chloride**
a chemical compound, used in producing some plastics, that is believed to be oncogenic.

■ **virga**
rain which fall in the sky but evapourates before it reaches the ground.

■ **virgin materials**
resources extracted from nature in their raw form, such as timber or metal ore.

■ **viscosity**
the molecular friction within a fluid that produces flow resistance.

■ **volatile**
any substance that evaporates readily.

■ **volatile liquids**
liquids which easily vaporise or evaporate at room temperature.

■ **volatile solids**
those solids in water or other liquids that are lost on ignition of the dry solids at 550° centigrade.

■ **volatile synthetic organic chemicals**
chemicals that tend to volatilise or evaporate.

■ **volume reduction**
processing waste materials to decrease the amount of space they occupy, usually by compacting, shredding, incineration, or composting.

■ **volumetric tank test**
one of several tests to determine the physical integrity of a storage tank; the volume of fluid in the tank is measured directly or calculated from product-level changes. A marked drop in volume indicates a leak.

■ **vulnerability analysis**
assessment of elements in the community that are susceptible to damage if hazardous materials are released.

vulnerable zone
an area over which the airborne concentration of a chemical accidentally released could reach the level of concern.

waste
1.unwanted materials left over from a manufacturing process. 2. Refuse from places of human or animal habitation.

waste characterisation
identification of chemical and microbiological constituents of a waste material.

waste exchange
arrangement in which companies exchange their wastes for the benefit of both parties.

waste feed
the continuous or intermittent flow of wastes into an incinerator.

waste generation
the weight or volume of materials and products that enter the waste stream before recycling, composting, landfilling, or combustion takes place. Also can represent the amount of waste generated by a given source or category of sources.

waste load allocation
1. The maximum load of pollutants each discharger of waste is allowed to release into a particular waterway. Discharge limits are usually required for each specific water quality criterion being, or expected to be, violated.
2. The portion of a stream's total assimilative capacity assigned to an individual discharge.

waste minimisation
measures or techniques that reduce the amount of wastes generated during industrial production processes; term is also applied to recycling and other efforts to reduce the amount of waste going into the waste stream.

waste piles
non-containerised, lined or unlined accumulations of solid, non-flowing waste.

waste reduction
using source reduction, recycling, or composting to prevent or reduce waste generation.

waste stream
the total flow of solid waste from homes, businesses, institutions, and manufacturing

plants that is recycled, burned, or disposed of in landfills, or segments thereof such as the "residential waste stream" or the "recyclable waste stream."

■ **waste treatment lagoon**
impoundment made by excavation or earth fill for biological treatment of wastewater.

■ **waste treatment plant**
a facility containing a series of tanks, screens, filters and other processes by which pollutants are removed from water.

■ **waste treatment stream**
the continuous movement of waste from generator to treater and disposer.

■ **waste-heat recovery**
recovering heat discharged as a by-product of one process to provide heat needed by a second process.

■ **waste-to-energy facility/municipal-waste combustor**
facility where recovered municipal solid waste is converted into a usable form of energy, usually via combustion.

■ **wastewater**
the spent or used water from a home, community, farm, or industry that contains dissolved or suspended matter. Water Pollution The presence in water of enough harmful or objectionable material to damage the water's quality.

■ **wastewater infrastructure**
the plan or network for the collection, treatment, and disposal of sewage in a community. The level of treatment will depend on the size of the community, the type of discharge, and/or the designated use of the receiving water.

■ **wastewater operations and maintenance**
actions taken after construction to ensure that facilities constructed to treat wastewater will be operated, maintained, and managed to reach prescribed effluent levels in an optimum manner.

■ **water cycle**
the recycling of water between the earth and the atmosphere.

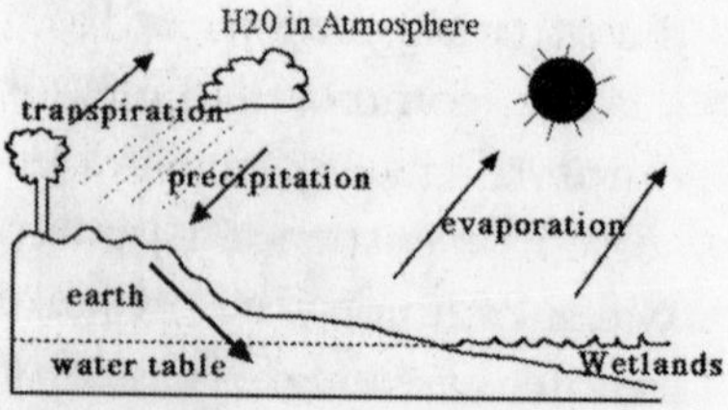

■ **water energy**
energy from moving water.

■ **water purveyor**
a public utility, mutual water company, county water district, or municipality that delivers drinking water to customers.

■ **water quality criteria**
levels of water quality expected to render a body of water suitable for its designated use. Criteria are based on specific levels of pollutants that would make the water harmful if used for drinking, swimming, farming, fish production, or industrial processes.

■ **water quality-based limitations**
effluent limitations applied to dischargers when mere technology-based limitations would cause violations of water quality standards. Usually applied to discharges into small streams.

■ **water quality-based permit**
a permit with an effluent limit more stringent than one based on technology performance. Such limits may be necessary to protect the designated use of receiving waters (e.g. recreation, irrigation, industry or water supply).

■ **water solubility**
the maximum possible concentration of a chemical compound dissolved in water. If a substance is water soluble it can very readily disperse through the environment.

■ **water storage pond**
an impound for liquid wastes designed to accomplish some degree of biochemical treatment.

■ **water supplier**
one who owns or operates a public water system.

■ **water supply system**
the collection, treatment, storage, and distribution of potable water from source to consumer.

■ **water table**
the level of groundwater.

■ **water treatment lagoon**
an impound for liquid wastes designed to accomplish some degree of biochemical treatment.

■ **water vapour**
the gas state of water.

■ **water well**
an excavation where the intended use is for location, acquisition, development, or artificial recharge of ground water.

■ **waterborne disease outbreak**
the significant occurence of acute illness associated with drinking water from a public water system that is deficient in treatment, as determined by appropriate local or state agencies.

■ **watershed**
the land area that drains into a stream; the watershed for a major river may encompass a number of smaller watersheds that ultimately combine at a common point.

■ **watershed approach**
a coordinated framework for environmental management that focuses public and private efforts on the highest priority problems within hydrologically-defined geographic areas taking into consideration both ground and surface water flow.

■ **watershed area**
a topographic area within a line drawn connecting the highest points uphill of a drinking water intake into which overland flow drains.

■ **water-soluble packaging**
packaging that dissolves in water; used to reduce exposure risks to pesticide mixers and loaders.

■ **water-source heat pump**
heat pump that uses wells or heat exchangers to transfer heat from water to the inside of a building. Most such units use ground water.

■ **weed**
any plant that grows where it is not wanted.

■ **weight of scientific evidence**
considerations in assessing the interpretation of pub-

lished information about toxicity—quality of testing methods, size and power of study design, consistency of results across studies, and biological plausibility of exposure-response relationships and statistical associations.

■ **weir**
1. a wall or plate placed in an open channel to measure the flow of water. 2. A wall or obstruction used to control flow from settling tanks and clarifiers to ensure a uniform flow rate and avoid short-circuiting.

■ **well**
a bored, drilled, or driven shaft, or a dug hole whose depth is greater than the largest surface dimension and whose purpose is to reach underground water supplies or oil, or to store or bury fluids below ground.

■ **well field**
area containing one or more wells that produce usable amounts of water or oil.

■ **well injection**
the subsurface emplacement of fluids into a well.

■ **well monitoring**
measurement by on-site instruments or laboratory methods of well water quality.

■ **well plug**
a watertight, gastight seal installed in a bore hole or well to prevent movement of fluids.

■ **well point**
a hollow vertical tube, rod, or pipe terminating in a perforated pointed shoe and fitted with a fine-mesh screen.

■ **wellhead protection area**
a protected surface and subsurface zone surrounding a well or well field supplying a public water system to keep contaminants from reaching the well water.

■ **wet deposits**
air pollutants that mix with moisture in the air before falling to the ground.

■ **wetlands**
an area that is saturated by surface or ground water with vegetation adapted for life under those soil conditions, as swamps, bogs, fens, marshes, and estuaries.

■ **wettability**
the relative degree to which a fluid will spread into or coat a solid surface in the presence of other immiscible fluids.

■ **wettable powder**
dry formulation that must be mixed with water or other liquid before it is applied.

■ **wheeling**
the transmission of electricity owned by one entity through the facilities owned by another (usually a utility).

■ **whole-effluent-toxicity tests**
tests to determine the toxicity levels of the total effluent from a single source as opposed to a series of tests for individual contaminants.

■ **wild flower**
a flowering plant that grows in woods, deserts, or other natural areas.

■ **wildlife refuge**
an area designated for the protection of wild animals, within which hunting and fishing are either prohibited or strictly controlled.

■ **wind breaks**
the practice of planting trees and shrubs to protect fields from soil erosion by wind.

■ **wind energy**
energy from moving air.

■ **wire-to-wire efficiency**
the efficiency of a pump and motor together.

■ **wood packaging**
wood products such as pallets, crates, and barrels.

■ **wood treatment facility**
an industrial facility that treats lumber and other wood products for outdoor use. The process employs chromated copper arsenate, which is regulated as a hazardous material.

■ **wood-burning-stove pollution**
air pollution caused by emissions of particulate matter, carbon monoxide, total suspended particulates, and polycyclic organic matter from wood-burning stoves.

■ **working level**
a unit of measure for documenting exposure to radon decay products, the so-called "daughters." One working

level is equal to approximately 200 picocuries per liter.

■ **working level month (WLM)**

a unit of measure used to determine cumulative exposure to radon.

■ **xenobiota**

any biotum displaced from its normal habitat; a chemical foreign to a biological systemYard Waste. The part of solid waste composed of grass clippings, leaves, twigs, branches, and other garden refuse.

■ **yellow-boy**

iron oxide flocculants (clumps of solids in waste or water); usually observed as orange-yellow deposits in surface streams with excess iron content.

■ **yield**

the quantity of water (expressed as a rate of flow or total quantity per year) that can be collected for a given use from surface or groundwater sources.

■ **zero air**

atmospheric air purified to contain less than 0.1 ppm total hydrocarbons.

■ **zoo**

a controlled environment where we visit and learn about animals.

■ **zooplankton**

tiny, free-floating organisms in aquatic systems. Unlike phytoplankton, zooplankton cannot produce their own food, and so are consumers.